Zong Woo Geem (Ed.)

Music-Inspired Harmony Search Algorithm

Studies in Computational Intelligence, Volume 191

Editor-in-Chief
Dr. Zong Woo Geem
Westat
1650 Research Blvd. TA1105
Rockville, Maryland 20850
USA
E-mail: Zwgeem@gmail.com

Further volumes of this series can be found on our homepage: springer.com

Vol. 169. Nadia Nedjah, Luiza de Macedo Mourelle and Janusz Kacprzyk (Eds.)
Innovative Applications in Data Mining, 2009
ISBN 978-3-540-88044-8

Vol. 170. Lakhmi C. Jain and Ngoc Thanh Nguyen (Eds.)
Knowledge Processing and Decision Making in Agent-Based Systems, 2009
ISBN 978-3-540-88048-6

Vol. 171. Chi-Keong Goh, Yew-Soon Ong and Kay Chen Tan (Eds.)
Multi-Objective Memetic Algorithms, 2009
ISBN 978-3-540-88050-9

Vol. 172. I-Hsien Ting and Hui-Ju Wu (Eds.)
Web Mining Applications in E-Commerce and E-Services, 2009
ISBN 978-3-540-88080-6

Vol. 173. Tobias Grosche
Computational Intelligence in Integrated Airline Scheduling, 2009
ISBN 978-3-540-89886-3

Vol. 174. Ajith Abraham, Rafael Falcón and Rafael Bello (Eds.)
Rough Set Theory: A True Landmark in Data Analysis, 2009
ISBN 978-3-540-89886-3

Vol. 175. Godfrey C. Onwubolu and Donald Davendra (Eds.)
Differential Evolution: A Handbook for Global Permutation-Based Combinatorial Optimization, 2009
ISBN 978-3-540-92150-9

Vol. 176. Beniamino Murgante, Giuseppe Borruso and Alessandra Lapucci (Eds.)
Geocomputation and Urban Planning, 2009
ISBN 978-3-540-89929-7

Vol. 177. Dikai Liu, Lingfeng Wang and Kay Chen Tan (Eds.)
Design and Control of Intelligent Robotic Systems, 2009
ISBN 978-3-540-89932-7

Vol. 178. Swagatam Das, Ajith Abraham and Amit Konar
Metaheuristic Clustering, 2009
ISBN 978-3-540-92172-1

Vol. 179. Mircea Gh. Negoita and Sorin Hintea
Bio-Inspired Technologies for the Hardware of Adaptive Systems, 2009
ISBN 978-3-540-76994-1

Vol. 180. Wojciech Mitkowski and Janusz Kacprzyk (Eds.)
Modelling Dynamics in Processes and Systems, 2009
ISBN 978-3-540-92202-5

Vol. 181. Georgios Miaoulis and Dimitri Plemenos (Eds.)
Intelligent Scene Modelling Information Systems, 2009
ISBN 978-3-540-92901-7

Vol. 182. Andrzej Bargiela and Witold Pedrycz (Eds.)
Human-Centric Information Processing Through Granular Modelling, 2009
ISBN 978-3-540-92915-4

Vol. 183. Marco A.C. Pacheco and Marley M.B.R. Vellasco (Eds.)
Intelligent Systems in Oil Field Development under Uncertainty, 2009
ISBN 978-3-540-92999-4

Vol. 184. Ljupco Kocarev, Zbigniew Galias and Shiguo Lian (Eds.)
Intelligent Computing Based on Chaos, 2009
ISBN 978-3-540-95971-7

Vol. 185. Anthony Brabazon and Michael O'Neill (Eds.)
Natural Computing in Computational Finance, 2009
ISBN 978-3-540-95973-1

Vol. 186. Chi-Keong Goh and Kay Chen Tan
Evolutionary Multi-objective Optimization in Uncertain Environments, 2009
ISBN 978-3-540-95975-5

Vol. 187. Mitsuo Gen, David Green, Osamu Katai, Bob McKay, Akira Namatame, Ruhul A. Sarker and Byoung-Tak Zhang (Eds.)
Intelligent and Evolutionary Systems, 2009
ISBN 978-3-540-95977-9

Vol. 188. Agustín Gutiérrez and Santiago Marco (Eds.)
Biologically Inspired Signal Processing for Chemical Sensing, 2009
ISBN 978-3-642-00175-8

Vol. 189. Sally McClean, Peter Millard, Elia El-Darzi and Chris Nugent (Eds.)
Intelligent Patient Management, 2009
ISBN 978-3-642-00178-9

Vol. 190. K.R. Venugopal, K.G. Srinivasa and L.M. Patnaik
Soft Computing for Data Mining Applications, 2009
ISBN 978-3-642-00192-5

Vol. 191. Zong Woo Geem (Ed.)
Music-Inspired Harmony Search Algorithm, 2009
ISBN 978-3-642-00184-0

Zong Woo Geem
(Ed.)

Music-Inspired Harmony Search Algorithm

Theory and Applications

Springer

Dr. Zong Woo Geem
Westat
1650 Research Blvd. TA1105
Rockville, Maryland 20850
USA
E-mail: Zwgeem@gmail.com

ISBN 978-3-642-00184-0 e-ISBN 978-3-642-00185-7

DOI 10.1007/978-3-642-00185-7

Studies in Computational Intelligence ISSN 1860949X

Library of Congress Control Number: 2008944108

Typeset & Cover Design: Scientific Publishing Services Pvt. Ltd., Chennai, India.

Printed in acid-free paper

9 8 7 6 5 4 3 2 1

springer.com

Preface

Calculus has been used in solving many scientific and engineering problems. For optimization problems, however, the differential calculus technique sometimes has a drawback when the objective function is step-wise, discontinuous, or multi-modal, or when decision variables are discrete rather than continuous. Thus, researchers have recently turned their interests into metaheuristic algorithms that have been inspired by natural phenomena such as evolution, animal behavior, or metallic annealing.

This book especially focuses on a music-inspired metaheuristic algorithm, harmony search. Interestingly, there exists an analogy between music and optimization: each musical instrument corresponds to each decision variable; musical note corresponds to variable value; and harmony corresponds to solution vector. Just like musicians in Jazz improvisation play notes randomly or based on experiences in order to find fantastic harmony, variables in the harmony search algorithm have random values or previously-memorized good values in order to find optimal solution.

The recently-developed harmony search algorithm has been vigorously applied to various optimization problems. Thus, the goal of this book is to show readers full spectrum of the algorithm in theory and applications in the form of an edited volume with the following subjects: justification as a metaheuristic algorithm by Yang; literature review by Ingram and Zhang; multi-modal approach by Gao, Wang & Ovaska; computer science applications by Mahdavi; engineering applications by Fesanghary; structural design by Saka; water and environmental applications by Geem, Tseng & Williams; groundwater modeling by Ayvaz; geotechnical analysis by Cheng; energy demand forecasting by Ceylan; sound classification in hearing aids by Alexandre, Cuadra & Gil-Pita; and therapeutic medical physics by Panchal.

As an editor of this book, I'd like to express my deepest thanks to reviewers and proofreaders including Mike Dreis, John Galuardi, Sanghun Kim, Una-May O'Reilly, Byungkyu Park, Ronald Wiles, and Ali Rıza Yıldız, as well as the above-mentioned chapter authors. Furthermore, as a first inventor of the harmony search algorithm, I especially thank Joel Donahue, Chung-Li Tseng, Joong Hoon Kim, and the late G. V. Loganathan (victim of Virginia Tech shooting) for their ideas and support. Finally, I'd like to share the joy of the publication with my family who are unceasing motivators in life.

Zong Woo Geem
Editor

Synopsis

Recently music-inspired harmony search algorithm has been proposed and vigorously applied to various scientific and engineering applications such as music composition, Sudoku puzzle solving, tour planning, web page clustering, structural design, water network design, vehicle routing, dam scheduling, ground water modeling, soil stability analysis, ecological conservation, energy system dispatch, heat exchanger design, transportation energy modeling, pumping operation, model parameter calibration, satellite heat pipe design, medical physics, etc.

However, these applications of the harmony search algorithm are dispersed in various journals, proceedings, degree theses, technical reports, books, and magazines, which makes readers difficult to draw a *big picture* of the algorithm. Thus, this book is designed to putting together all the latest developments and cutting-edge studies of theoretical backgrounds and practical applications of the harmony search algorithm for the first time, in order for readers to efficiently understand a full spectrum of the algorithm and to foster new breakthroughs in their fields using the algorithm.

Contents

Harmony Search as a Metaheuristic Algorithm

Xin-She Yang

Department of Engineering, University of Cambridge, Trumpington Street,
Cambridge CB2 1PZ, UK
xy227@cam.ac.uk

Abstract. This first chapter intends to review and analyze the powerful new Harmony Search (HS) algorithm in the context of metaheuristic algorithms. We will first outline the fundamental steps of HS, and show how it works. We then try to identify the characteristics of metaheuristics and analyze why HS is a good metaheuristic algorithm. We then review briefly other popular metaheuristics such as particle swarm optimization so as to find their similarities and differences with HS. Finally, we will discuss the ways to improve and develop new variants of HS, and make suggestions for further research including open questions.

Keywords: Harmony Search, Metaheuristic Algorithms, Diversification, Intensification, Optimization.

1 Introduction

When listening to a beautiful piece of classical music, who has ever wondered if there is any connection between playing music and finding an optimal solution to a tough design problem such as the water network design or other problems in engineering? Now for the first time ever, scientists have found such an interesting connection by developing a new algorithm, called Harmony Search. HS was first developed by Geem et al. in 2001 [1]. Though it is a relatively new metaheuristic algorithm, its effectiveness and advantages have been demonstrated in various applications. Since its first appearance in 2001, it has been applied to many optimization problems including function optimization, engineering optimization, design of water distribution networks, groundwater modeling, energy-saving dispatch, truss design, vehicle routing, and others [2, 3]. The possibility of combining harmony search with other algorithms such as Particle Swarm Optimization has also been investigated.

Harmony search is a music-based metaheuristic optimization algorithm. It was inspired by the observation that the aim of music is to search for a perfect state of harmony. The effort to find the harmony in music is analogous to find the optimality in an optimization process. In other words, a jazz musician's improvisation process can be compared to the search process in optimization. On one hand, the perfectly pleasing harmony is determined by the audio aesthetic standard. A musician always intends to produce a piece of music with perfect harmony. On the other hand, an optimal solution to an optimization problem should be the best solution available to the problem under the given objectives and limited by constraints. Both processes intend to produce the best or optimum.

Such similarities between two processes can be used to develop a new algorithm by learning from each other. Harmony Search is just such a successful example by transforming the qualitative improvisation process into quantitative optimization

Z.W. Geem (Ed.): Music-Inspired Harmony Search Algorithm, SCI 191, pp. 1–14.
springerlink.com

process with some idealized rules, and thus turning the beauty and harmony of music into a solution for various optimization problems.

2 Harmony Search as a Metaheuristic Method

Before we introduce the fundamentals of the HS algorithm, let us first briefly describe the way to describe the aesthetic quality of music. Then, we will discuss the pseudo code of the HS algorithm and two simple examples to demonstrate how it works.

2.1 Aesthetic Quality of Music

The aesthetic quality of a musical instrument is essentially determined by its pitch (or frequency), timbre (or sound quality), and amplitude (or loudness). Timbre is largely determined by the harmonic content that is in turn determined by the waveforms or modulations of the sound signal. However, the harmonics that it can generate will largely depend on the pitch or frequency range of the particular instrument.

Different notes have different frequencies. For example, the note A above middle C (or standard concert A4) has a fundamental frequency of $f_0=440$ Hz. As the speed of sound in dry air is about $v=331+0.6T$ m/s (where T is the temperature in degrees Celsius), the A4 note has a wavelength $\lambda=v/f_0 \approx 0.7795$ m at room temperature T=20 ℃. When we adjust the pitch, we are in fact trying to change the frequency. In music theory, pitch p_n in MIDI is often represented as a numerical scale (a linear pitch space) using the following formula

$$p_n = 69 + 12\log_2(\frac{f}{440Hz}), \tag{1}$$

or

$$f = 440 \times 2^{(p_n - 69)/12}, \tag{2}$$

which means that the A4 notes has a pitch number 69. On this scale, octaves correspond to size 12 while semitone corresponds to size 1, which leads to the fact that the ratio of frequencies of two notes that are an octave apart is 2:1. Thus, the frequency of a note is doubled (halved) when it raised (lowered) an octave. For example, A2 has a frequency of 110Hz while A5 has a frequency of 880Hz.

The measurement of harmony where different pitches occur simultaneously, like any aesthetic quality, is subjective to some extent. However, it is possible to use some standard estimation for harmony. The frequency ratio, pioneered by ancient Greek mathematician Pythagoras, is a good way for such estimation. For example, the octave with a ratio of 1:2 sounds pleasant when playing together, so are the notes with a ratio of 2:3. However, it is unlikely for any random notes played by a monkey to produce a pleasant harmony.

2.2 Harmony Search

In order to explain the Harmony Search in more detail, let us first idealize the improvisation process by a skilled musician. When a musician is improvising, he or she has three possible choices: (1) playing any famous tune exactly from his or her memory; (2) playing something similar to the aforementioned tune (thus adjusting the pitch slightly); or (3) composing new or random notes. Geem et al. formalized these three options into quantitative optimization process in 2001, and the three corresponding components become: usage of harmony memory, pitch adjusting, and randomization [1].

The usage of harmony memory (HM) is important because it ensures that good harmonies are considered as elements of new solution vectors. In order to use this memory effectively, the HS algorithm adopts a parameter $r_{accept} \in [0,1]$, called harmony memory considering (or accepting) rate. If this rate is too low, only few elite harmonies are selected and it may converge too slowly. If this rate is extremely high (near 1), the pitches in the harmony memory are mostly used, and other ones are not explored well, leading not into good solutions. Therefore, typically, we use r_{accept}=0.7 ~ 0.95.

The second component is the pitch adjustment which has parameters such as pitch bandwidth b_{range} and pitch adjusting rate r_{pa}. As the pitch adjustment in music means changing the frequency, it means generating a slightly different value in the HS algorithm [1]. In theory, the pitch can be adjusted linearly or nonlinearly, but in practice, linear adjustment is used. So we have

$$x_{new} = x_{old} + b_{range} \times \varepsilon \tag{3}$$

where x_{old} is the existing pitch stored in the harmony memory, and x_{new} is the new pitch after the pitch adjusting action. This action produces a new pitch by adding small random amount to the existing pitch [2]. Here ε is a random number from uniform distribution with the range of [-1, 1]. Pitch adjustment is similar to the mutation operator in genetic algorithms. We can assign a pitch-adjusting rate (r_{pa}) to control the degree of the adjustment. A low pitch adjusting rate with a narrow bandwidth can slow down the convergence of HS because of the limitation in the exploration of only a small subspace of the whole search space. On the other hand, a very high pitch-adjusting rate with a wide bandwidth may cause the solution to scatter around some potential optima as in a random search. Thus, we usually use r_{pa}=0.1~0.5 in most applications.

Harmony Search

---

begin
  Define objective function $f(\mathbf{x})$, $\mathbf{x}=(x_1,x_2, \ldots,x_d)^T$
  Define harmony memory accepting rate (r_{accept})
  Define pitch adjusting rate (r_{pa}) *and other parameters*
  Generate Harmony Memory with random harmonies
  while *(t<max number of iterations)*
    while *(i<=number of variables)*
      if *(rand<*r_{accept}*), Choose a value from HM for the variable i*
        if *(rand<*r_{pa}*), Adjust the value by adding certain amount*
        end if
      else *Choose a random value*
      end if
    end while
    Accept the new harmony (solution) if better
  end while
  Find the current best solution
end

Fig. 1. Pseudo code of the Harmony Search algorithm

The third component is the randomization, which is to increase the diversity of the solutions. Although the pitch adjustment has a similar role, it is limited to certain area and thus corresponds to a local search. The use of randomization can drive the system further to explore various diverse solutions so as to attain the global optimality.

These three components in harmony search can be summarized as the pseudo code shown in Figure 1. In this pseudo code, we can see that the probability of randomization is

$$P_{random} = 1 - r_{accept}, \tag{4}$$

and the actual probability of the pitch adjustment is

$$P_{pitch} = r_{accept} \times r_{pa}. \tag{5}$$

2.3 Implementation

The above-mentioned three components of the HS algorithm can be easily implemented using any programming language. However, Matlab gives more straightforward way because it also provides visualization as shown in Figure 2.

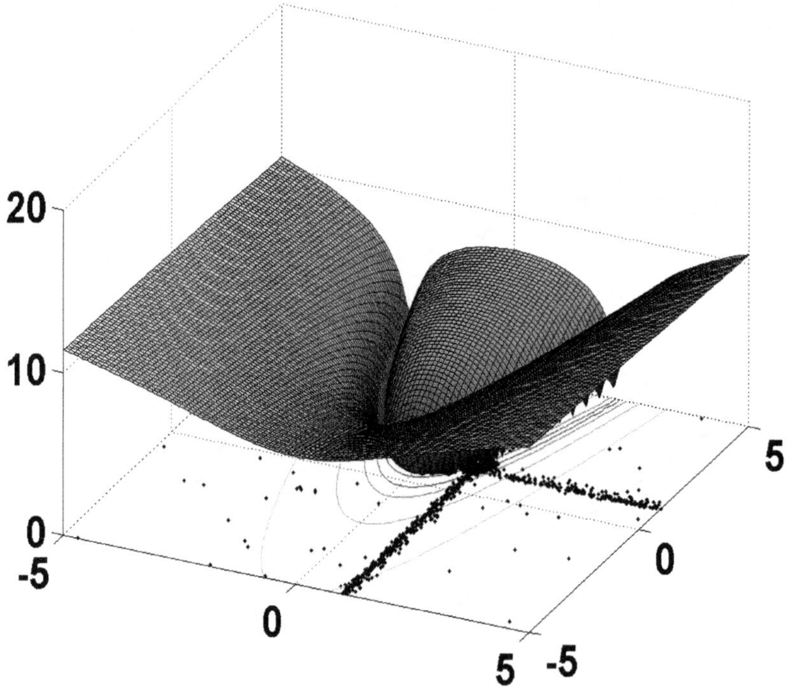

Fig. 2. The search paths of finding the global optimal solution (1,1) using the harmony search

For the first benchmark example, we test Rosenbrock's logarithmic banana function as follows:

$$f(x,y) = \ln[1 + (1-x)^2 + 100(y - x^2)^2] \tag{6}$$

where $(x, y) \in [-10, 10] \times [-10, 10]$ and the global minimum $f_{min}=0$ at (1, 1). The HS algorithm found the global optimum successfully. The variations of these solutions and their paths are shown in Figure 2.

We have used 20 harmonies with harmony accepting rate r_{accept}=0.95, and pitch adjusting rate r_{pa}=0.7. The search paths are plotted together with the landscape of $f(x,y)$. From Figure 2, we can see that the pitch adjustment is more intensive in local regions (two thin strips).

As a further example, we present Michalewicz's bivariate function as follows:

$$f(x,y) = -\sin(x)\sin^{20}\left(\frac{x^2}{\pi}\right) - \sin(y)\sin^{20}\left(\frac{2y^2}{\pi}\right), \tag{7}$$

where a global minimum $f_{min} \approx -1.801$ at [2.20319, 1.57049] in the domain $0 \le x \le \pi$ and $0 \le y \le \pi$. This global minimum was found by the HS algorithm as shown in Figure 3.

In addition to the above-mentioned two benchmark examples, this book contains many successful examples of the HS algorithm in solving various tough optimization problems, and also provides comparison among the HS algorithm and other ones. Such a comparison among different types of algorithms is still an area of active research.

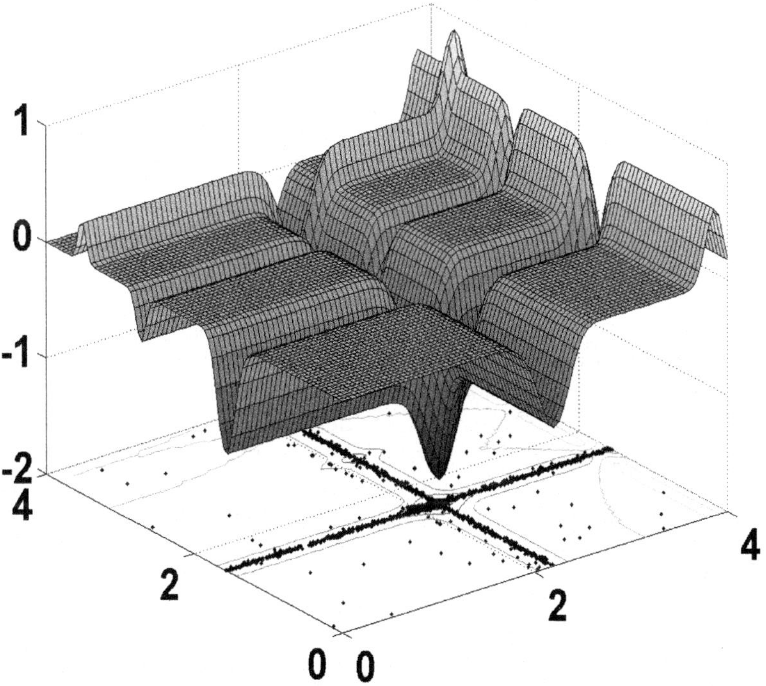

Fig. 3. Harmony search for Michalewicz's bivariate function

3 Other Metaheuristics

3.1 Metaheuristics

Heuristic algorithms typically intend to find a good solution to an optimization problem by 'trial-and-error' in a reasonable amount of computing time. Here 'heuristic' means to 'find' or 'search' by trials and errors. There is no guarantee to find the best or optimal solution, though it might find a better or improved solution than an educated guess. Broadly speaking, heuristic methods are local search methods because their searches focus on the local variations, and the optimal or best solution can locate outside of this local region. However, a high-quality feasible solution in the local region of interest is usually accepted as a good solution in many optimization problems in practice if time is the major constraint.

Metaheuristic algorithms are advanced heuristic algorithms. Because '*meta-*' means 'beyond' or 'higher-level', metaheuristic literally means to find the solution using higher-level techniques, though certain trial-and-error processes are still used. Broadly speaking, metaheuristics are considered as higher-level techniques or strategies that intend to combine lower-level techniques and tactics for exploration and exploitation of the huge solution space. In recent years, the word '*metaheuristics*' refers to all modern higher-level algorithms, including Evolutionary Algorithms (EA) including Genetic Algorithms (GA), Simulated Annealing (SA), Tabu Search (TS), Ant Colony Optimization (ACO), Particle Swarm Optimization (PSO), Bee Algorithms (BA), Firefly Algorithms (FA), and certainly Harmony Search [4].

There are two important components in modern metaheuristics: intensification and diversification. Such terminologies are derived from Tabu search [5]. For an algorithm to be efficient and effective, it must be able to generate a diverse range of solutions including the potentially optimal solutions so as to explore the whole search space effectively, while it intensifies its search around the neibourhood of an optimal or nearly optimal solution. In order to do so, every part of the search space must be accessible though not necessarily visited during the search. Diversification is often in the form of randomization with a random component attached to a deterministic component in order to explore the search space effectively and efficiently, while intensification is the exploitation of past solutions so as to select the potentially good solutions via elitism or use of memory or both [4-6].

Any successful metaheuristic algorithm requires a good balance between these two important, seemingly opposite, components [6]. If the intensification is too strong, only a fraction of solution space might be visited, and there is a risk of being trapped in a local optimum. It is often the case of gradient-based techniques such as Newton-Raphson method. Meanwhile, if the diversification is too strong, the algorithms converge too slowly because solutions jump around some potentially optimal solutions. Typically, the solutions start with some randomly generated (or educated guess) solutions, and gradually reduce their diversification while increasing their intensification at the same time.

Another important feature of modern metaheuristics is that an algorithm is either trajectory-based or population-based. For example, simulated annealing is a good example of trajectory-based algorithm because the path of the active search point (or agent) forms a Brownian motion-like trajectory with its movement towards some attractors. On the other hand, genetic algorithm is a good example of population-based method since the solution search is carried out by multiple genes or agents in parallel. It is difficult to decide which type of method is more efficient as both types work almost equally successfully under appropriate conditions. There are some hints from the recent studies that population-based algorithms might be more efficient for multiobjective multimodal optimization problems as multiple search actions are in parallel. This might be true from the implementation viewpoint. However, it is far from conclusive and there exists virtually no theoretical research to back this up.

3.2 Popular Metaheuristic Algorithms

We will now briefly introduce some popular metaheuristic algorithms, and we will try to see how intensification and diversification are used.

3.2.1 Evolutionary Algorithms

Evolutionary algorithms are the name for a subset of evolutionary computation [7]. They are the search methods that were inspired from Charles Darwin's natural selection and survival of the fittest. Evolutionary algorithms are population-based search algorithms and they use genetic operators to a certain degree. These operators typically include crossover or reproductive recombination, mutation, inheritance and selection based upon their fitness.

Genetic algorithms are by far the most popular and widely used [8, 9]. However, other evolutionary search methods, including genetic programming (GP), evolutionary strategies (ES), and evolutionary programming (EP), are also popular. Briefly speaking, genetic programming is an evolutionary machine-learning technique in the framework of genetic algorithms where each individual in the population is a computer program using a scheme-style computer language such as Lisp. The objective is to find the optimal computer program to perform a user-defined task.

Evolutionary strategy is another class of nature-inspired evolutionary optimization techniques. It mainly uses mutation, selection and element-wise average for intermediate recombination as the genetic operators. It has been applied to a wide range of optimization problems.

Evolutionary programming uses arbitrary data structures and representations tailored to suit a particular problem domain, and they are combined with the essence of genetic algorithms so as to solve generalized complex optimization problems. EP is very similar to ES, but does not have the recombination operator. The main difference between EP and other methods is that EP does not use exchange of string segments, and thus there is no crossover or recombination between individuals. The primary genetic operator is mutation, often using Gaussian mutation for real-valued functions. However, its modern variants are more diverse.

Here we will briefly introduce the basic idea of genetic algorithms. Genetic algorithms were developed by John Holland in the 1960s and 1970s [8], using crossover, mutation and selection for adaptive, optimization and other problems. The essence of genetic algorithms involves the encoding of an optimization function or a set of functions into arrays of binary or character strings to represent chromosomes, and the manipulation of these strings by genetic operators with the ultimate aim to find the optimal solution. The encoding is often carried out into multiple binary strings (called population) though real-valued encoding and representations are more often used in modern applications. The initial population then evolves by generating new-generation individuals via crossover of two randomly selected parent strings, and/or the mutation of some random bits. Whether a new individual is selected or not is based on its fitness, which is linked in some way to the objective function.

The advantages of the genetic algorithms over traditional optimization algorithms are the ability of dealing with complex problems and parallelism. GA search methods can deal with various types of optimization, whether the objective functions are stationary or non-stationary (time-dependent), linear or nonlinear, continuous or discontinuous, or even with random noise. As individuals in a population act like independent agents, each subset of the population can explore the search space in many directions simultaneously. This feature makes it ideal for the parallel implementation of the genetic algorithms.

3.2.2 Simulated Annealing

Simulated annealing is probably the best example of modern metaheuristic algorithms. It was first developed by Kirkpatrick et al. in 1983 [10], inspired by the annealing process of metals during heat treatment and also by Metropolis algorithms for Monte Carlo simulations. The basic idea of the SA algorithm is similar to dropping a bouncing ball over a landscape. As the ball bounces and looses energy, it will settle down at certain local minimum. If the ball is allowed to bounce enough time and loose energy slowly enough, the ball may eventually fall into the globally lowest location, and hence the minimum will be reached. Of course, we can use not only single ball (standard simulated annealing) but also multiple balls (parallel simulated annealing).

The optimization process typically starts from an initial guess with higher energy. It then moves to another location randomly with slightly reduced energy. The move is accepted if the new state has lower energy and the solution improves with a better objective or lower value of the objective function for minimization. However, even if the solution is not improved, it is still accepted with a probability of

$$p = \exp[-\frac{\delta E}{kT}], \tag{8}$$

which is a Boltzmann-type probability distribution. Here, T is the temperature of the system and k is the Boltzmann constant which can be set into one for simplicity. The energy difference δE is often related to the objective function $f(\boldsymbol{x})$ to be optimized. The trajectory in the simulated annealing is a piecewise path, and this is virtually a Markov chain because the new state (new solution) depends solely on current state/solution and transition probability p.

Here the diversification via randomization produces new solutions (locations). Whether the new solution is accepted or not is determined by the probability. If T is too low ($T \to 0$), then any $\delta E > 0$ (worse solution) will rarely be accepted as $p \to 0$, and the diversity of the solutions is subsequently limited. On the other hand, if T is too high, the system is at a high-energy state and most new changes will be accepted. However, the minima are not easily reached. Thus, the temperature T is an essential controlling parameter for the balance of diversification and intensification.

3.2.3 Ant Colony Optimization

Another population-based metaheuristic algorithm is the ant colony optimization (ACO) which was first formulated by Dorigo and further developed by other pioneers [11-13]. This algorithm was based upon the characteristics of behaviours of social ants. For discrete routing and scheduling problems, multiple ants are often used. Each virtual ant will preferably choose a route covered with higher pheromone concentration, and it also deposits more pheromone at the same time. If there is no previously deposited pheromone, then each ant will move randomly. In addition, the pheromone concentration will decrease gradually due to the evaporation, often with a constant evaporation rate.

Here the diversification of the solutions is represented by the randomness and the choice probability of agents along a specific route. The intensification is implicitly manipulated by the pheromone concentration and the evaporation rate. However, the evaporation rate can also affect the diversification in some way. This algorithm is exceptionally successful in the combinatorial optimization problems. Again, some fine balance between the diversification and intensification is needed to ensure a faster and efficient convergence and to ensure the quality of the solutions.

3.2.4 Particle Swarm Optimization

As genetic algorithm is population-based metaheuristics and simulated annealing is a trajectory-based one, we now introduce another population-based metaheuristic algorithm, named particle swarm optimization. The PSO was developed by Kennedy and Eberhart [14], inspired by the swarm behaviour of fish and bird schooling in nature. Unlike the single trajectory scheme used in simulated annealing, this algorithm searches the solution space by adjusting multiple trajectories of individual agents (called particles). The motion of the particles has two major components: a stochastic component and a deterministic component in terms of velocity and position vectors (solution vectors)

$$v_i^{t+1} = v_i^t + \alpha\, \varepsilon_1 (x_i - g^*) + \beta\, \varepsilon_2 (x_i - x_i^*), \quad x_i^{t+1} = x_i^t + v_i^t, \tag{9}$$

where v_i^t and x_i^t are the velocity and position of particle i at time t, respectively. ε_1 and ε_2 are two random vectors, while α and β are constants (often called the learning parameters). The diversification is controlled by the combination of random vectors and learning parameters.

The intensification is mainly represented by the deterministic motion towards the updated current best x_i^t for particle i, and the current global best g^* for all particles. As the particles approach to the optima, their motion and randomness are reduced. There are many different variants of PSO in the literature [4,14].

There is a hidden or implicit feature in the PSO algorithm, that is the broadcasting ability of the current global best g^* to other particles. That can be thought as either the use of memory or some higher-level strategy so as to speed up the convergence and explore the search space more effectively and efficiently. If the diversification (or learning) parameter is large, a larger part of the search space will be explored; however, it will converge more slowly. On the other hand, high-level intensification will make the algorithm converge quickly, but not necessarily to the right solution set.

3.2.5 Firefly Algorithm

The fascinating flashing light of fireflies in the tropical summer can be used to develop interesting nature-inspired metaheuristic algorithms for optimization. The Firefly Algorithm was developed by Xin-She Yang [4], based on the idealization of the flashing characteristics of fireflies. There are three major components in the FA optimization: 1) A firefly will be attracted to more brighter or more attractive fireflies, and at the same time they will move randomly; 2) the attractiveness is proportional to the brightness of the flashing light which will decrease with distance, therefore, the attractiveness will be evaluated in the eye of the beholders (other fireflies); 3) The decrease of light intensity is controlled by the light absorption coefficient γ which is in turn linked to a characteristic scale.

The new solution is generated by

$$x_i^{t+1} = x_i^t + \alpha\, \varepsilon_1 + \beta \exp[-\gamma\, r_{ij}^2]\, \varepsilon_2 (x_i - x_j), \tag{10}$$

where r_{ij} is the distance, not necessarily the physical distance, between two fireflies i and j. In addition, FA is obviously a population-based algorithm, which may share many similarities with particle swarm optimization. In fact, it has been proved by Yang [4] that when $\gamma \to \infty$, the firefly algorithm will become an accelerated version of PSO, while $\gamma \to 0$, the FA reduces to a version of random search algorithms.

In the FA optimization, the diversification is represented by the random movement component, while the intensification is implicitly controlled by the attraction of different fireflies and the attractiveness strength β. Unlike other metaheuristics, the interaction between exploration and exploitation is intermingled in some way; this might be an important factor for its success in solving multiobjective and multimodal optimization problems.

Obviously, there are many other metaheuristic algorithms that are currently used, including, but not limited to, tabu search, cross-entropy, scatter search, cultural algorithm, flog leaping algorithm, artificial immune system, artificial bee algorithms, photosynthetic algorithm, enzyme algorithm, etc [15-20]. As we will discuss later, the hybridization of diversification and intensification components is a useful technique to develop new algorithms.

4 Characteristics of HS and Comparison

After the brief introduction to other metaheuristic algorithms, we are now ready to analyze the similarities and differences of the Harmony Search algorithm in the general context of metaheuristics.

4.1 Diversification and Intensification

In reviewing other metaheuristic algorithms, we have repetitively focused on two major components: diversification and intensification. They are also referred to as exploration and exploitation [6]. These two components are seemingly contradicting each other, but their balanced combination is crucially important to the success of any metaheuristic algorithms [4, 6].

Proper diversification or exploration makes sure that the search in solution space can explore as many locations and regions as possible in an efficient and effective manner. It also ensures that the evolving system will not be trapped in biased local optima. Diversification is often represented in the implementation as the randomization and/or additional stochastic component superposed onto the deterministic components. If the diversification is too strong, it may explore too many locations in a stochastic manner, and subsequently will slow down the convergence of the algorithm; if the diversification is too weak, there is a risk that the solution space explored is so limited and the solutions are biased and trapped in local optima, or even lead to meaningless solutions.

On the other hand, the appropriate intensification or exploitation intends to exploit the history and experience of the search process. It aims to ensure to speed up the convergence when necessary by reducing the randomness and limiting diversification. Intensification is often carried out by using memory such as in tabu search and/or elitism such as in genetic algorithm. In other algorithms, it is much more elaborate to use intensification such as the case in simulated annealing and firefly algorithms. If the intensification is too strong, it could result in premature convergence, leading to biased local optima or even meaningless solutions because the search space is not well explored. On the contrary, if the intensification is too weak, convergence becomes slow.

The optimal balance of diversification and intensification is required, and such a balance itself is an optimization process. Fine-tuning of parameters is often required to improve the efficiency of an algorithm for a particular problem. There is No Free Lunch in any optimization problem [21]. A substantial amount of studies might be required to choose the right algorithm for the right optimization problem [16], though a systematic guidance lacks for such a choice.

4.2 Why HS Is Successful

Now if we analyze the Harmony Search algorithm in the context of the major components of metaheuristics and try to compare with other metaheuristic algorithms, we can identify its ways of handling intensification and diversification, and probably understand why it is a very successful metaheuristic algorithm.

In the HS algorithm, diversification is essentially controlled by the pitch adjustment and randomization -- here there are two subcomponents for diversification, which might be an important factor for the high efficiency of the HS method. The first subcomponent of playing a new pitch (or generating a new value) via randomization would be at least at the same level of efficiency as in other algorithms that handle randomization. However, an additional subcomponent for HS diversification is the pitch adjustment operation performed with the probability of r_{pa}. Pitch adjustment is carried out by tuning the pitch within a given bandwidth. A small random amount is added to or subtracted from an existing pitch (or solution) stored in HM. Essentially, pitch adjustment is a refinement process of local solutions. Both memory consideration and pitch adjustment ensure that good local solutions are retained while the randomization makes the algorithm to explore global search space effectively. The subtlety is the fact that HS operates controlled diversification around good solutions, and intensification as well. The randomization explores the search space more widely and efficiently;

while the pitch adjustment ensures that the newly generated solution is good enough, or not too far from existing good solutions.

The intensification in the HS algorithm is represented by the harmony memory accepting rate r_{accept}. A high harmony acceptance rate means that good solutions from the history/memory are more likely to be selected or inherited. This is equivalent to a certain degree of elitism. Obviously, if the acceptance rate is too low, solutions will converge more slowly. As mentioned earlier, this intensification is enhanced by the controlled pitch adjustment. Such interactions between various components could be another important factor for the success of the HS algorithm over other algorithms, as it will be demonstrated again in other chapters of this book.

In addition, the structure of the HS algorithm is relatively easier. This advantage makes it very versatile to combine HS with other metaheuristic algorithms [22]. For algorithm parameters, there are some evidences to suggest that HS is less sensitive to chosen parameters, which means that we may not have to fine-tune these parameters to get quality solutions.

Furthermore, the HS algorithm is a population-based metaheuristic, which means that a group of multiple harmonies can be used in parallel. Proper parallelism usually leads to better performance with higher efficiency. The good combination of parallelism with elitism as well as a fine balance of intensification and diversification is the key to the success of the HS algorithm.

5 Further Research

The power and efficiency of the HS algorithm seem obvious after discussion and comparison with other metaheuristics; however, there are some unanswered questions concerning the whole class of the algorithm. Currently, the HS algorithm like other popular metaheuristics works well under appropriate conditions. However, we usually do not fully understand why and how they work so well. For example, when choosing the harmony accepting rate, we usually use a higher value, say, 0.7 to 0.95. This value is determined by experience or inspiration from genetic algorithm where the mutation rate should be low, and thus the accepting rate of the existing gene components should be high. However, it is very difficult to say what range of values and which combinations are surely better than others.

In general, there lacks a theoretical framework for metaheuristics to provide some analytical guidance to the following important issues: How to improve the efficiency for a given problem? What conditions are needed to guarantee a good rate of convergence? How to prove the global optima are reached for the given metaheuristic algorithm? These are still open questions that need further research. The encouraging thing is that many researchers are interested in tackling these difficult challenges, and important progress has been made concerning the convergence of certain algorithm (SA). Any progress concerning the convergence of HS and other algorithms would be influentially profound.

Even without a solid framework, this does not discourage scientists to develop more variants and/or hybrid algorithms. In fact, the algorithm development itself is a metaheuristic process in a similar manner to the key components of HS: using existing successful algorithms; developing slightly different variants based on the existing algorithms; and formulating completely new metaheuristic algorithms. In using the existing algorithms, we have to identify the right algorithm for the right problem. Often, we have to change and reformulate the problem slightly and/or improve the algorithm slightly in order to find the solutions more efficiently. Sometimes, we have to develop a new algorithm from scratch to solve a tough optimization problem.

There are many ways to develop new algorithms. From the metaheuristic viewpoint, the most heuristic way is probably to develop a new algorithm by hybridization. That is to say, a new algorithm can be made based on the right combination of existing metaheuristic algorithms. For example, by combining a trajectory-type simulated annealing with multiple agents, the parallel simulated annealing can be developed. In the context of the HS algorithm, by the combination of HS with PSO, the global-best harmony search has been developed [22].

As in the case of any efficient metaheuristic algorithms, the most difficult thing is probably to find the right or optimal balance between diversity and intensity in searching the solutions. Here, the most challenging task in developing new hybrid algorithms is probably to find out the right combination of which feature/components of existing algorithms.

A further extension of the HS algorithm will be to solve multiobjective problems more naturally and more efficiently. At the moment, most of the existing studies, though very complex and tough *per se*, have mainly focused on the optimization with a single objective with or without a few criteria. The next challenges would be to use the HS algorithm to solve tough multiobjective and multicriteria NP-hard optimization problems.

Whatever the challenges will be, more HS algorithms will be applied to various problems and more systematic studies will be performed for the analysis of HS mechanism. Also, more hybrid algorithms based on HS will be developed in the future.

References

1. Geem, Z.W., Kim, J.H., Loganathan, G.V.: A new heuristic optimization algorithm: Harmony search. Simulation 76, 60–68 (2001)
2. Lee, K.S., Geem, Z.W.: A new meta-heuristic algorithm for continuous engineering optimization: harmony search theory and practice. Comput. Methods Appl. Mech. Engrg. 194, 3902–3933 (2005)
3. Harmony Search Algorithm (2007) (accessed December 7, 2008), `http://www.hydroteq.com`
4. Yang, X.S.: Nature-inspired metaheuristic algorithms. Luniver Press (2008)
5. Glover, F., Laguna, M.: Tabu search. Kluwer Academic Publishers, Dordrecht (1997)
6. Blum, C., Roli, A.: Metaheuristics in combinatorial optimization: Overview and conceptual comparison. ACM Comput. Surv. 35, 268–308 (2003)
7. De Jong, K.: Evolutionary computation: a unified approach. MIT Press, Cambridge (2006)
8. Holland, J.H.: Adaptation in natural and artificial systems. The University of Michigan Press, Ann Arbor (1975)
9. Goldberg, D.E.: Genetic algorithms in search, optimization, and machine learning. Addison Wesley, Reading (1989)
10. Kirkpatrick, S., Gelatt, C.D., Vecchi, M.P.: Optimization by simulated annealing. Science 220, 671–680 (1983)

11. Dorigo, M., Stutzle, T.: Ant colony optimization. MIT Press, Cambridge (2004)
12. Bonabeau, E., Dorigo, M., Theraulaz, G.: Swarm intelligence: from natural to artificial systems. Oxford University Press, Oxford (1999)
13. Dorigo, M., Blum, C.: Ant colony optimization theory: a survey. Theor. Comput. Sci. 344, 243–278 (2005)
14. Kennedy, J., Eberhart, R.C.: Particle swarm optimization. In: Proceedings of IEEE Int. Conf. Neural Networks, pp. 1942–1948 (1995)
15. Yang, X.S.: Biology-derived algorithms in engineering optimization. In: Olarius, S., Zomaya, A. (eds.) Handbook of Bioinspired Algorithms and Applications. Chapman & Hall/CRC, Boca Raton (2005)
16. Yang, X.S.: Mathematical optimization: from linear programming to metaheuristics. Cambridge Int. Science Publishing, UK (2008)
17. Engelbrecht, A.P.: Fundamentals of computational swarm intelligence. Wiley, Chichester (2005)
18. Perelman, L., Ostfeld, A.: An adaptive heuristic cross-entropy algorithm for optimal design of water distribution systems. Engineering Optimization 39, 413–428 (2007)
19. Karaboga, D., Basturk, B.: On the performance of artificial bee colony (ABC) algorithm. Applied Soft Computing 8, 687–697 (2008)
20. Yang, X.S.: New enzyme algorithm, Tikhonov regularization and inverse parabolic analysis. In: Simos, T., Maroulis, G. (eds.) Advances in Computational Methods in Science and Engineering – ICCMSE 2005, vol. 4, pp. 1880–1883 (2005)
21. Wolpert, D.H., Macready, W.G.: No free lunch theorems for optimization. IEEE Transaction on Evolutionary Computation 1, 67–82 (1997)
22. Omran, M., Mahdavi: Global-best harmony search. Applied Math. Computation 198, 643–656 (2008)

Overview of Applications and Developments in the Harmony Search Algorithm

Gordon Ingram and Tonghua Zhang

Department of Chemical Engineering, Curtin University of Technology,
Perth, Australia
{g.ingram,t.zhang}@curtin.edu.au

Abstract. The Harmony Search (HS) algorithm appeared around the year 2000 and it now has a substantial literature. The aim of this chapter is to inform the reader of the diversity of HS applications and the range of modified and hybrid HS methods that have been developed. The chapter contains five sections. In the first, the different types of optimisation problem are described. Section two provides an overview of the growth in the literature, a chronology of some HS highlights and a breakdown of HS work by discipline. In the third section, HS applications in nine discipline areas are reported. The fourth section classifies the types of modifications that have been made to the original HS method and summarises the innovations in many of the modified algorithms. Lastly, we take a step back and reflect on the current state of the HS literature.

Keywords: Harmony Search, Literature Review, Chronology, Industrial Application, Algorithm Development.

1 Introduction

Since the initial development of the Harmony Search (HS) algorithm by Zong Woo Geem in 2000 [1], HS methods have been applied to a diverse range of problems—from structural design to solving Sudoku puzzles, from musical composition to medical imaging, from heat exchanger design to course timetabling. This chapter presents an extensive, though not exhaustive, summary of HS applications and associated developments in the HS method itself.

After some preliminaries in Section 1.1 and 1.2, we treat the HS literature in three ways to cater for readers with different interests. For those wishing to see the broad view of HS, Section 2 provides an overview of HS activities, including the historical growth in the HS literature, a chronology of selected HS 'highlights', and a summary of its application areas. For readers interested in a particular field, such as water distribution or information technology, we summarise HS applications by discipline area in Section 3. For researchers concerned with the methods of metaheuristic optimisation, Section 4 outlines the many modifications that have been proposed to the original HS method. Section 5 briefly reflects on the body of HS work to date.

1.1 Mathematical Description of Optimisation Problems

HS is an optimisation method, and it is worthwhile clarifying at the outset the nature of the problem being addressed. The *unconstrained*, *single objective*, *multivariable* optimisation problem may be written as

$$\min_{\mathbf{x}} f(\mathbf{x}) \tag{1}$$

Z.W. Geem (Ed.): Music-Inspired Harmony Search Algorithm, SCI 191, pp. 15–37.
springerlink.com

where f is the *objective function* to be minimised by varying a vector of *decision*, or *design, variables* $\mathbf{x} = (x_1, \ldots, x_n)^T$. *Single variable* optimisation corresponds to $n = 1$.

In some cases, all the decision variables are *continuous*, being described by lower and upper bounds, $x_i \in \Re : x_i^L \leq x_i \leq x_i^U$. In other cases the variables are *discrete*, able to take on only particular distinct values, $x_i \in X_i = \{X_{i,1}, \ldots, X_{i,k}\}$, which includes the special case of binary variables: $x_i \in \{0,1\}$. *Mixed variable* problems have both discrete and continuous variables in $\mathbf{x}$. The objective function f itself may be continuous or discontinuous.

Many practical problems involve *constrained optimisation* where $\mathbf{x}$ needs to satisfy additional equality and/or inequality relationships:

$$h_i(\mathbf{x}) = 0; \quad i = 1, \ldots, p ; \tag{2}$$

$$g_i(\mathbf{x}) \geq 0; \quad i = 1, \ldots, q . \tag{3}$$

Multi-objective optimisation aims to minimise several objective functions simultaneously:

$$\min_{\mathbf{x}} f_i(\mathbf{x}); \quad i = 1, \ldots, r . \tag{4}$$

HS has been used to solve all of these different types of optimisation problem.

1.2 The Motivation for Undertaking Optimisation

Table 1 shows some common reasons for conducting optimisation studies. This serves to clarify some of the applications discussed later in this chapter. More detailed examples may be found in the following chapters of this book.

Table 1. Why is Optimisation Performed?

Role of Optimisation	Description
Benchmarking	Some optimisation problems have become de facto standards that are used to compare existing optimisation methods and test new ones
Design	This is the process of selecting materials, configurations, sizes and conditions for a man-made system that best satisfy some design requirements
Calibration or parameter estimation	Mathematical models of physical systems often contain parameters that need to be adjusted to optimise the fit between the model predictions and real-world data
Scheduling	These problems involve finding the best sequence of events or tasks to be allocated to resources or equipment, typically to minimise total production time or cost
Route finding	In any network, such as a road system or the Internet, there is a need to find the best way to transport items between different locations
Analysis	Systems tend to seek a state of lowest energy, which is a form of natural optimisation—to help understand these systems we can set up and solve optimisation problems

2 Overview of HS Activities

2.1 Growth in the Literature

Previous reviews of the HS literature have focused on applications in civil engineering [2], the range of industrial applications [3], HS in the context other metaheuristic optimisation algorithms [4, 5], and HS methods for chemical engineering [6]. Figure 1 shows that there has certainly been sustained and increasing interest in HS methods since their appearance in 2000. Up to 2004 the annual number of publications was modest, but from 2005 onwards there is a marked increase in HS activity.

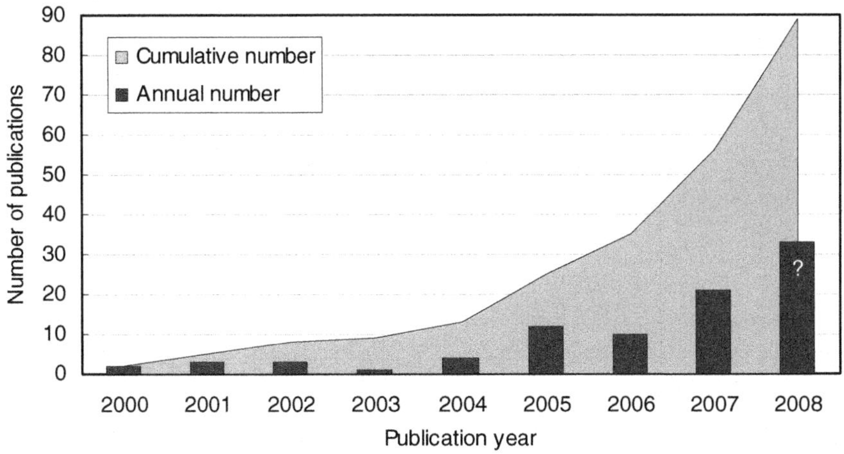

Fig. 1. Indicative Growth in the HS Literature to November 2008

2.2 Historical Highlights

The initial development of the method was conducted during Geem's PhD studies in the Department of Civil and Environmental Engineering at Korea University [1]. While the study covered benchmark optimisation, parameter estimation, and the Travelling Salesman Problem (TSP), most effort was devoted to the design of water distribution networks. Indeed, water distribution and water-related studies remain a strong theme in the HS literature. Geem, Kim and Loganathan presented the HS method to the wider community in 2001 [7]. These and subsequent selected HS applications and innovations are presented as a chronology in Figure 2.

Major studies into HS optimisation for structural design were reported in 2004 for continuous design variables [8] and in 2005 for discrete variables [9]. These papers, along with [4] and [10], are doubly significant because they introduce flowcharts for the HS algorithm. The flowcharts also depict constraint handling.

In 2005 a study into the stability analysis of soil slopes was reported using a hybrid HS and Genetic Algorithm (GA) method [11]. Soil stability analysis has continued as a key topic for HS applications, with most work originating in China.

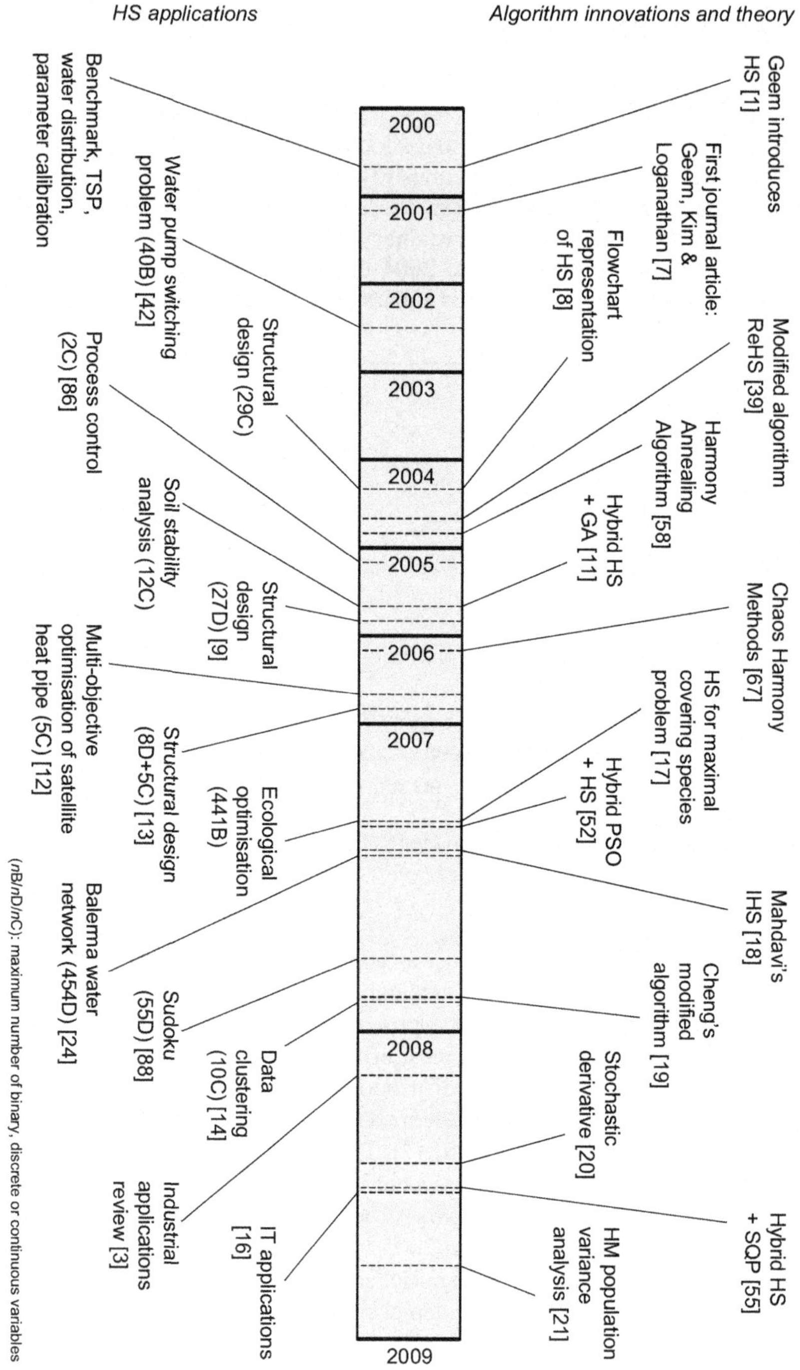

Fig. 2. Chronology of Selected Developments and Applications of the HS Algorithm to November 2008

The first multi-objective optimisation problem appears to have been tackled in 2006 [12] for the design and operation of a heat pipe on a satellite. One of the few mixed variable problems was also reported in 2006 [13].

Applications related to Information Technology (IT), including data clustering [14, 15] and information routing in computer networks [16], have begun appearing in recent years.

Since the introduction of the original HS algorithm, researchers have been investigating ways to improve its performance or adapt it to new types of problems. Even the two original studies [1, 7] present the basic method and two modifications. Figure 2 shows only a few of the many innovations that have been reported. Extensive modifications to the original algorithm were made to solve a type of ecological conservation problem for Oregon, USA, termed the Maximal Covering Species Problem (MCSP) [17]. Mahdavi, Fesanghary and Damangir [18] developed the Improved Harmony Search (IHS), which has been used in several subsequent studies. The highly modified algorithm by Cheng, Li and Chi [19] is another noteworthy development. Around 45% of studies to date (Figure 1) use the original HS algorithm as presented by Yang in Section 2.2 of Chapter 1, with the remainder using modified versions.

Modified methods have also been developed by hybridising HS with other metaheuristic optimisation methods, such as GA or Particle Swarm Optimisation (PSO). Approximately half of the modified HS algorithms may be classified as hybrid methods. Section 4 discusses hybrid and other modified HS algorithms.

A little theoretical analysis of the HS algorithm has been conducted. Studies include HS convergence [1, 7], the 'stochastic partial derivative' [20] and a population variance analysis for Harmony Memory [21].

Some of the larger problems studied to date using original or modified HS methods include

- MCSP for ecological conservation with 441 binary variables [17], and a pipe network layout problem with 112 binary variables [22, 23];
- Balerma water distribution network with 454 discrete variables [24], and university course timetabling with 450 discrete variables [25];
- Benchmark optimisation problems with up to 100 continuous variables [26, 27], and soil stability analysis with up to 71 continuous variables [28];
- In terms of mixed variable optimisation, structural optimisation problems with up to 8 discrete and 13 continuous variables [29], and 8 discrete and 5 continuous variables [13].

2.3 Areas of Activity

Until around 2005 about half the published HS studies were devoted to the design of water distribution networks. Benchmark optimisation, structural design and route finding problems comprised the remaining portion of the literature. Since 2005 or so the range of applications areas has expanded. Figure 3 shows the current approximate distribution of HS applications, as measured by the number of publications devoted to that topic. Section 3 goes through each of these discipline areas, describing the typical problems addressed and citing specific publications.

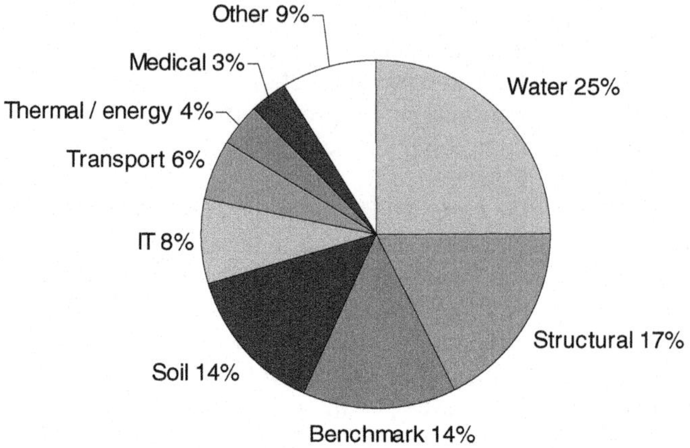

Fig. 3. Approximate Breakdown of HS Applications by Discipline Area as of November 2008

3 Applications of HS by Discipline Area

Many applications, excluding benchmark problems, have been summarised in [3]. A significant portion of publications compare the performance of HS with other optimisation algorithms; some also report the sensitivity of the results to the parameters of the HS algorithm. More detail on specific issues in each discipline area is provided in the other chapters of this book.

3.1 Water-Related Applications

Several studies have focussed on the design of municipal water distribution networks, in particular the selection of optimal pipe diameters [1, 7, 10, 24, 30, 31]. Brief results are also reported in [20, 32–34]. Typically there is one fixed-pressure supply node in the network and many demand nodes, with the structure (connectivity) of the distribution network being given. The elevations and distances between nodes (pipe lengths) are also specified. The aim is to select the diameter for each pipe segment that minimises the total capital cost of the network, given that pipes come in certain standard diameters and smaller-diameter pipes cost less. The problem is constrained by the laws of fluid mechanics—mass and energy conservation equations—and there are also customer requirements at each demand node for some minimum water pressure. A hydraulic simulator, such as EPANET, is used to perform the fluid mechanics calculations. The networks considered range in size from the 8-pipe 'two-loop' network [10] to the 454-pipe Balerma network in Spain [24]. A related study is the design of coffer dam drainage pipes by [35].

Geem [36] considered the optimal sizing of both the pipes and the water supply pump in a distribution network. The objective was to minimise the total cost, comprising capital costs for the pump and pipes, plus the operating cost for the pump.

The optimal structure (layout) of rectilinear branched pipe networks has been studied for 3×3 and 8×8 regular grids of nodes [22, 23]. This problem has one supply node with the rest of the grid points being demand nodes. The decision is whether or not to place a pipe between adjacent nodes. No loops were permitted.

The location of a leak in a pipeline can be estimated from the transient pressure or flow rate response to the sudden closure of a downstream valve. Leak detection was investigated using HS and GA in [37, 38]. The rehabilitation of pipe networks was studied in [39, 40] with a modified HS method. The problem was to select the year and type of rehabilitation work undertaken to minimise the total cost over the network's lifetime.

A water pump switching problem was investigated in [41] and briefly in [42]. Ten pump stations, each containing four pumps in series, were located along a pipeline. The problem was to decide which pumps should operate to minimise the energy cost subject to constraints on the pump suction and discharge pressures. The optimal scheduling of a system of four dams was studied in [43]. The aim was to find the water release schedule that maximised both hydropower and irrigation benefits. There were constraints on the dam outflows, and instantaneous and final storage levels.

Several water-related studies involve parameter estimation. Parameters for the nonlinear Muskingum model, which can be used in the prediction of volumetric water flow rates for flood routing, were fitted in [1, 44] and also discussed in [4, 32]. Paik and co-workers [45] developed a rainfall-runoff model that required estimation of 16–40 continuous parameters. Ayvaz [14] combined HS with fuzzy *C*-means clustering to estimate the zone structure and zonal transmissivities for a heterogeneous aquifer.

3.2 Structural Design

Typical structural design problems involve selecting the best cross-sectional areas or designation codes for beams making up a structure to minimise the structure's weight. The configuration and lengths of beams forming the structure and any applied external loads are specified. There are constraints on the maximum tensile and compressive stresses that beams should experience, and on the maximum deflections of selected nodes. Both continuous variable and discrete variable problems have been studied. Considerable use is made of symmetry in the structure to reduce the number of decision variables required.

Continuous variable applications using basic HS include two-dimensional (plane) trusses with 10–200 bars, three-dimensional (space) trusses with 22–72 bars, and a 120-bar dome truss [8]. Discrete variable problems, namely a 52-bar plane truss, 47-bar power line tower, and 25- and 72-bar space trusses, were reported in [9]. Saka [46] presented optimised designs for geodesic domes having three to six rings (75–285 bars). Erdal and Saka [47] optimised 24–264-member grillages, which are frameworks of longitudinal and transverse beams used to support platforms. Degertekin [48, 49] reported designs for several frame structures, including a 1-bay, 8-storey plane frame and a 4-storey, 84-bar space frame. Brief results on structural optimisation are also reported in [4, 50]. A hybrid PSO-HS algorithm was used to optimise various structures: 10- and 17-bar plane trusses, and 22-, 25- and 72-bar space trusses [51–53].

A few structural design studies go beyond optimisation of beam cross-sections. Both cross-sectional area or designation and nodal positions were optimised for an 18-bar plane truss in [4] and a 25-bar space truss in [13]. Lo [29] reported an extensive study using a modified HS method for several combinations of different types of design variables, including cross-sectional area, nodal position and topology (the presence or absence of a beam). Structures included 10- and 18-bar plane trusses, 25- and 39-bar space trusses, 52- and 132-bar dome trusses, and a 160-bar three-dimensional transmission tower. Lo's modified algorithm, HS-DLM, focussed on constraint handling using Discrete Lagrange Multipliers (DLM).

Lee and Geem [4] reported on the geometrical optimisation of a pressure vessel and a welded beam. These examples are also considered by Jang et al [54] using a new hybrid Nelder-Mead (NM) simplex algorithm – HS technique, and by Mahdavi et al [18] using an adaptive harmony search method (IHS). A modified algorithm combining gradient-based Sequential Quadratic Programming (SQP) and HS was applied to three structural problems, including the welded beam, in [55]. The design of an offshore mooring system was considered in [56]. The sizes of the mooring system components were sought to minimise the system cost, subject to constraints on the displacement of the moored vessel, cable tension, and length of mooring chain lying on the seabed.

3.3 Benchmark Optimisation

Twelve benchmark continuous optimisation problems, including unconstrained and constrained examples with 2–10 variables, were investigated using the original HS algorithm in an extensive study in [4]. Benchmark problems are also briefly discussed in [1, 7–9, 20, 30, 42].

Omran and Mahdavi [27] compared the performance of the original HS, IHS [18] and a new Global-best HS (GHS) algorithm on ten continuous functions of up dimension 100, and six integer programming problems with up to 30 variables. They also studied the sensitivity to HS parameters and the effect of noise. Mukhopadhyay et al [21] compared another modified HS method with IHS, GHS and Differential Evolution (DE) for five continuous test functions. Gao et al [57] developed a modified HS method specifically for multi-modal functions and applied it to three 2-dimensional multi-modal functions.

Optimisation of a selection of continuous functions with 2–30 variables using the hybrid Harmony Annealing Algorithm (HAA) was reported in [58, 59]. The Rastrigin, Griewank and Sphere functions were optimised for 30, 50 and 100 variables using a hybrid PSO-HS algorithm developed for high-dimension problems in [26]. Results for the same three functions in 2–30 dimensions were reported in [60] for a GA-HS algorithm. A hybrid HS-DE method for uni-modal problems was tested on eight 50-dimensional benchmark functions in [57]. Jang et al [54] optimised two unconstrained and three constrained bench-mark functions with 2–7 continuous variables using a hybrid NM-HS method. A hybrid algorithm that combined elements from GA, HS, NM and the Tabu Search (TS) was tested on six continuous functions of dimension 2 to 10 in [61]. Fesanghary et al [55] applied a modified HS-SQP method to two constrained and two unconstrained benchmark problems.

3.4 Soil Stability Analysis

A body of soil with an inclined surface may become unstable and slip. The aim of slope stability analysis is to predict the location of the surface inside the soil body where slippage may occur (the critical slip surface) and to estimate the associated factor of safety, which is the ratio of inherent shear strength of the soil to the shear stress that it experiences. Several soil layers may be present with different properties. The location of slip surface is found by minimising the factor of safety.

Two-dimensional slip surfaces of arbitrary shape were analysed using HS in [62], modified HS methods in [63, 64], a hybrid GA-HS algorithm in [11, 65, 66], three new Chaos HS algorithms in [67] and hybrid PSO-HS in [28]. A hybrid PSO-HS method was also used for three-dimensional slope stability analysis in [68]. Cheng et al [19] presented a comprehensive comparison of six meta-heuristic optimisation methods for slope stability analysis, including two HS-based methods. These modified HS methods are discussed in more detail in [69, 70].

3.5 Information Technology Applications

Several IT applications tackle data clustering. The aim of clustering is to divide a data set into groups such that there is a high level of similarity for members within a group, but a low level of similarity between different groups. Clustering of web documents was studied in [71, 15] using three novel hybrid HS – *K*-means clustering algorithms. Fuzzy classification of the Fisher Iris data set was investigated in [72]. Initial classification was performed using the Fuzzy *C*-Means (FCM) method then optimisation of the fuzzy membership functions was accomplished by a new hybrid HS – Clonal Selection Algorithm (CSA) method. Malaki and co-workers [73] developed two hybrid IHS-FCM clustering algorithms, which were tested on a ~58,000 element NASA radiator data set.

Forsati et al [16, 74] studied multi-cast routing, which refers to the transmission of the data from a sender through a network to multiple recipients. Two modified HS algorithms were developed to solve the least cost multi-cast routing problem subject to maximum bandwidth and delay time constraints.

Cruz et al [75] used a new hybrid optimisation algorithm inspired by GA, Simulated Annealing (SA) and HS for a parameter estimation problem in the development of virtual urban environments.

3.6 Transport-Related Problems

A 20-city TSP was solved in [1, 7] and a 51-city TSP was also presented in [1], both using a modified HS algorithm. A school bus routing problem was investigated in [76, 77] and briefly in [32]. The problem was to find the required number of school buses and the best route for each bus subject to constraints on the bus seating capacity and maximum allowable journey time. A generalised orienteering problem, for the best touring in China, was solved using a modified HS method in [78]. The aim was to find the tour route that maximised the collective tourism opportunities offered by each city, subject to a constraint on the maximum tour length.

A parameter estimation problem for the annual energy demand of the Turkish transportation sector was reported in [79].

3.7 Thermal and Energy Applications

Geem and Hwangbo [12] performed multi-objective optimisation for the design of a satellite heat pipe. The problem involved finding the heat pipe dimensions and operating temperature to minimise the heat pipe's mass while maximising its thermal conductance.

Fesanghary et al [80] compared IHS and GA for the optimal design of a shell and tube heat exchanger, which included both annualised capital and operating costs.

The optimal operation of a system of cogeneration (combined heat and power) plants was studied in [81]. The aim was to determine the heat and power generation rates at each plant to minimise the total cost, subject to constraints on the total heat and power demand, and the feasible operating region for each plant.

3.8 Medical Studies

Some digital hearing aids can classify their acoustic environment and adapt the sounds transmitted to the wearer accordingly. Amor et al [82] used HS to determine the best subset from a set of 74 standard features of the acoustic environment to be used in the sound classifier.

High dose rate brachytherapy uses the temporary placement of tiny radiation sources into tissue using catheters for the treatment of cancer. Panchal [83] reported on the HS optimisation of the dwell time, which is the duration that the radiation source is left in the vicinity of the affected tissue. Dong et al [84] studied the detection of abnormalities in biological tissue from digital images using an adaptive-parameter HS algorithm.

3.9 Other Applications

Liu and Feng [85] used HS for system identification. They estimated ten discrete-valued parameters of a Controlled Auto-Regressive Moving Average (CARMA) model for oil well heat wash data. Two applications in process control have been reported: nonlinear model predictive control for set-point tracking using HAA [86], and synchronization of discrete-time chaotic systems using another modified HS method [87].

Further applications include the solution of Sudoku puzzles [88], musical composition [89], timetabling and room allocation for university courses [25], optimization of a milling process [90], and the selection of land parcels for ecological conservation (a MCSP) [17].

4 Developments in the HS Method

This section first presents a classification for the modifications made to the original HS method, and then it shows which modified algorithms use which particular innovations. Lastly, developments in the mathematical analysis of the HS method are reported.

4.1 Classification of Modifications to the Original HS

The original HS algorithm was presented in Chapter 1, and we briefly recap the parameters and steps in the algorithm here.

Harmony Memory (HM) is the principal HS data structure. It is a matrix that stores in its rows a selection of the current best harmonies (solution vectors):

$$\mathbf{HM} = \begin{bmatrix} \mathbf{x}^1 \\ \vdots \\ \mathbf{x}^{HMS} \end{bmatrix} \in \Re^{HMS \times n} \tag{5}$$

where HMS is the Harmony Memory Size. As shown in Figure 4, the steps in the HS algorithm are (i) initialising harmony memory; (ii) improvising a new harmony, that is, generating a new candidate solution vector; (iii) updating HM with the new harmony if appropriate; and (iv) returning to step (ii) until some termination criteria are satisfied. The improvisation of new harmonies in step (ii) involves three operations: random playing, harmony memory considering, and pitch adjusting, as described in Chapter 1. Several fixed-value parameters control these operations: the Harmony Memory Considering Rate (HMCR); Pitch Adjusting Rate (PAR); and either the bandwidth (b) for continuous variables or the neighbouring index (m) for discrete variables, which are both used in pitch adjusting. If the new harmony is better than the worst one currently in HM then HM is updated in step (iii) by replacing the poorest solution with the new harmony. The algorithm terminates when it reaches a certain maximum number of improvisations ($MaxImp$).

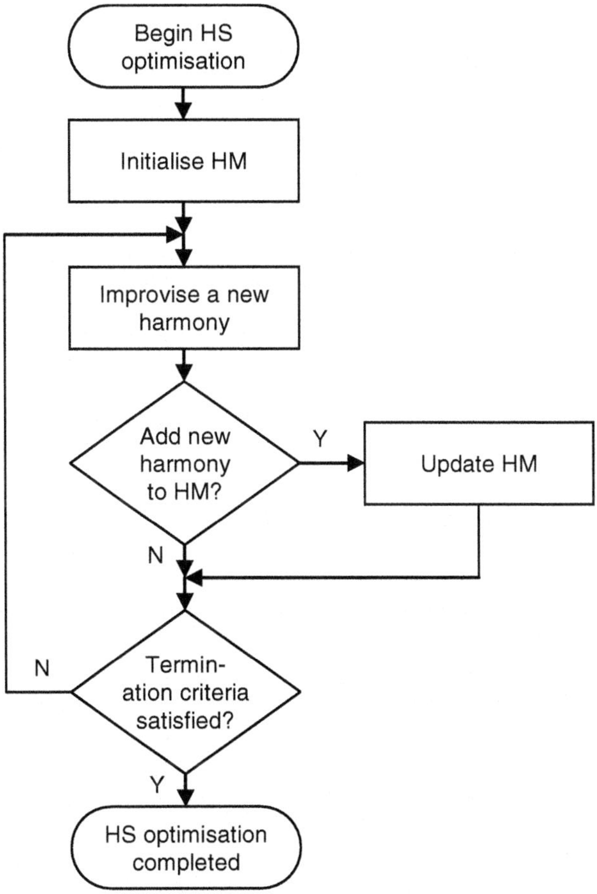

Fig. 4. Flowchart for the Original HS Algorithm

Modifications to the original algorithm may be classified into several categories, which are placed in context by Figure 4:

- Alternative initialisation procedures for HM, or an extended HM structure
  - *Example:* Degertekin [48] generated $2 \times HMS$ initial harmonies but placed only the best HMS of these into the initial HM;
- Variable, rather than fixed, parameters used when improvising a new harmony
  - *Example:* Mahdavi et al [18] advocated parameters for pitch adjusting that vary with the improvisation number $j = 1, \ldots, MaxImp$:

$$PAR(j) = PAR_{min} + \left(PAR_{max} - PAR_{min}\right) \times \frac{j}{MaxImp} \tag{6}$$

$$b(j) = b_{max} \exp\left[\ln\left(\frac{b_{min}}{b_{max}}\right)\frac{j}{MaxImp}\right]; \tag{7}$$

- New or revised operations for new harmony improvisation
  - *Example:* Li et al [64] introduced a non-uniform mutation operation from GA:

$$x'_{new,i} = \begin{cases} x_{new,i} + \Delta\left(j, x_i^U - x_{new,i}\right) & if\ r_1 \le 0.5 \\ x_{new,i} - \Delta\left(j, x_{new,i} - x_i^L\right) & if\ r_1 > 0.5 \end{cases} \tag{8}$$

  - where $\Delta(j, y) = y \times r_2 \times \left(1 - \frac{j}{MaxImp}\right)^b$ and $r_1, r_2 \in [0,1]$ are random numbers;
- Options for handling constraints during generation of new harmonies
  - *Example:* After generating a new harmony, Erdal [47] used two methods to handle constraints: if the new harmony was strongly infeasible, it was simply discarded; if the error was small, it was considered for inclusion in HM, but the acceptable error decreased as iterations progressed;
- Different criteria for deciding when to include a new harmony in HM
  - *Example:* Gao et al [57] included a new harmony only if it met three conditions: (i) it is better than the worst harmony in HM, and (ii) there are less than a critical number of similar harmonies already in HM, and (iii) its fitness is better than the average fitness of the similar harmonies;
- Revised termination criteria
  - *Example:* Cheng et al [19] terminated iterations when the best objective function value changed less than a small amount after a given number of iterations;
- Modifications to the algorithm's structure, that is, adding or removing blocks and changing the processing sequence in the flowchart (Figure 4)
  - Structural changes may be relatively small, for example generating multiple harmonies per improvisation [67], or they may be extreme, such as the HPSO method [52], which is essentially PSO with occasional use of HM considering to fix infeasible solutions.

Table 2. Algorithm–Innovation Matrix for Modified, Non-Hybrid HS Algorithms

Author's Algorithm Name	Parameters	Operations	HM	Constraint Handling	Termination Criteria	Structural Changes	Reference
HS	$\mathbf{x}^L$ & $\mathbf{x}^U$ tightened at *MaxImp*/2 for benchmark problem	City-inverting & neighbouring-city-going for TSP	-	-	-	-	[1]
Revised HS (ReHS)	*HMCR* & *PAR* are not fixed	-	-	-	-	-	[39, 40]
HS	-	-	-	Feasible solutions only stored in HM	-	-	[4, 8, 9]
Harmony Annealing Algorithm (HAA)	-	Random playing replaced by very fast SA	-	-	-	-	[58, 59, 86]
HS	*PAR* different for $m = \pm 1,2,3$ (PAR_1, PAR_2, PAR_3)	-	-	-	-	-	[78]
Modified HS (MHS)	-	Special training of elite harmonies; proportionate selection operator from GA (discarded)	Identical harmonies in HM replaced with new harmonies	-	-	-	[45]
Improved HS	-	Ensemble consideration for correlated variables	-	-	-	-	[33]
HS	-	-	Identical harmonies in HM replaced with new harmonies	Feasible solutions only stored in HM	-	-	[23]

Table 2. *(continued)*

Chaos HS Method (three versions)	$\mathbf{x}^L$ & $\mathbf{x}^U$ adjusted for chaotic harmonies	Chaotic harmony generation via logistic equation	-	-	-	Multiple harmonies generated per improvisation	[67]
Modified HS Method	-	HM considering and random playing modified	-	If solution is infeasible, a new solution is generated using a penalty or correcting strategy	-	All solutions may be pitch adjusted	[63]
Modified Harmony Method (MHM)	$\mathbf{x}^L$ & $\mathbf{x}^U$ tightened in final round of iterations	HM considering uses biased selection	All solutions in HM except best one are replaced several times during iterations	-	Based on lack of change in *f* and on number of improvisations	All solutions may be pitch adjusted; multiple harmonies generated per improvisation	[19, 69, 70]
Modified HS for MCSP	-	Random playing has biased probability; deterministic selection of land parcel every 10 iterations if species not already covered	HM initialisation with 5×*HMS* random vectors; at most two identical vectors allowed in HM	-	-	New harmony generation may stop if the maximum number of parcels is reached	[17]
HS	-	Random playing replaced with non-uniform mutation operator from GA	-	-	-	-	[64]
Improved HS (IHS)	*PAR* linearly increased and *b* exponentially decreased with iterations	-	-	-	-	-	[18, 80, 90]

Table 2. *(continued)*

HS	-	HM considering biased towards selecting courses with smallest number of available positions	HM initialisation only with feasible solutions	Feasibility maintained for new harmony generation	-	-	[25]
HS	-	-	HM initialisation with 2×*HMS* random vectors [48]; no identical vectors allowed in HM	-	Based on lack of change in *f* and on number of improvisations	-	[48, 49]
Adaptive HS (AHS)	*HMCR* linearly increased & *PAR*s linearly decreased with iterations; *PAR* different for $m = \pm1,2,3$	-	-	-	-	-	[84]
HS	-	-	-	Constraints handled using rejection & adaptive error strategies	-	-	[47]
Hybridizing HS Algorithm (HHSA)	As for IHS [18]	SQP used with small probability at each improvisation	-	-	-	At *MaxImp*, SQP used to polish each solution in HM	[55]
HSPR & HSNPI	*PAR* linearly increased with iterations (IHS); *PAR* different for $m = \pm1,2,3$	HM considering selects most frequent solution in HM (HSNPI)	Initialisation with valid solutions only (spanning trees)	-	-	-	[16, 74]

Table 2. *(continued)*

Modified HS for Multi-modal Problems	-	-	New harmony included in HM only if passes both fitness and dissimilarity tests	-	-	-	[57]
Hybrid Fuzzy HS Clustering (FHSClust) – FCM Algorithm	As for IHS [18]	-	-	-	-	At *MaxImp*, solution is polished with FCM	[73]
Modified HS	b = standard deviation of variable in HM	-	-	-	-	-	[21]
Global-best HS (GHS)	As for IHS [18]	Modified pitch adjusting using global best concept from PSO	-	-	-	-	[27]
Improved HS	PAR varies with j-according to grade concept from Dispersed PSO	-	-	-	-	-	[87]

In some modified algorithms only a single, simple change to the original HS method is made; in others several changes are combined in a complex way. Many individual innovations were summarised in [6].

4.2 Survey of Modified Algorithms

In this section, the innovations used in the current range of modified HS algorithms are outlined. We distinguish between hybrid and non-hybrid algorithms. We have defined a hybrid algorithm loosely as one whose structure is essentially no longer HS. Table 2 presents an algorithm–innovation matrix for modified, non-hybrid HS methods. Table 3 lists hybrid HS methods, and the reader is guided to the references to explore those methods further.

Table 3. Hybrid HS Methods

Author's Algorithm Name	Type of Hybrid	Reference
GA incorporated with Harmony Procedure	GA + HS	[11]
Hybrid Algorithm (HA)	GA + HS + NM + TS	[61]
Improved GA	GA + HS	[65, 66]
Mixed Search Algorithm (MSA)	PSO + HS	[28]
Novel Hybrid Real-Valued GA (NHRVGA)	GA + HS	[60]
Novel Hybrid PSO (NHPSO)	PSO + HS	[26]
Improved PSO (IPSO), Heuristic PSO (HPSO)	PSO + HS	[51–53]
New Hybrid Metaheuristic Algorithm	GA + SA + HS	[75]
HSCLUST / HClust, HKClust, IHKClust	HS / IHS + *K*-means	[15, 71]
Fusion of HS and DE (HS-DE)	HS + DE	[57]
Simplex-Harmony Search (SHS)	HS + NM	[54]
New version of PSO	PSO + HS	[68]
HS-DLM Algorithm	HS + DLM	[29]
Hybrid Optimization Method	CSA + HS	[72]

4.3 Theoretical Analyses of the Original HS Algorithm

A few publications attempt a theoretical analysis of the HS algorithm, although many present limited sensitivity analyses using the algorithm parameters HMS, HMCR and PAR.

In brief early work, Geem [1, 7] calculated the probability of finding the optimal solution in HM for $PAR = 0$. More recently, he derived the 'stochastic partial derivative' [20], which in fact estimates the probability that a particular value will be chosen for a discrete decision variable given the current contents of HM. A benchmark function and a water distribution problem were used to show how the 'derivative' evolves to favour the optimal solution. Mukhopadhyay and colleagues [21] present an informative population variance analysis for HS. They develop an expression for the expected variance of the solution vectors stored in HM, and use it to show how the HS algorithm can be modified to maintain diversification during its iterations.

5 Reflection on the HS Literature

It is an exciting time to be involved in the HS world. The literature is still relatively small, but is growing rapidly as seen in Figure 1. A critical mass of researchers is forming around the world. HS is expanding its horizons. From its original application to civil engineering and benchmark problems, there are now studies in medicine, process engineering, information technology, ecology and other disciplines. Fittingly, the Harmony Search has been applied to musical composition and hearing aid design! Surely the diversity of applications will continue grow as the algorithm becomes more widely known.

More than half the HS publications to date use some modification of the original algorithm. This chapter, and indeed this book, may catalyse further development of the method. In Chapter 1, Yang encouraged us to use heuristics to develop variant and hybrid HS algorithms. Mukhopadhyay and colleagues have shown us how a theoretical approach can lead to algorithmic innovations. The authors hope that the classification and summary in Section 4 may also help to inspire readers to further development of the HS method. There are intriguing hints that the power of the original HS lays somewhat untapped, let alone that of the modified algorithms.

References

1. Geem, Z.W.: Optimal design of water distribution networks using harmony search. PhD thesis, Korea University (2000)
2. Geem, Z.W., Geem, W.B.: Cutting-edge optimization technique and its applications to the civil engineering. The Magazine of the Korean Society of Civil Engineers 55, 340–349 (2007) (in Korean)
3. Geem, Z.W.: Harmony search applications in industry. In: Prasad, B. (ed.) Soft computing applications in industry. Springer, Berlin (2008)
4. Lee, K.S., Geem, Z.W.: A new meta-heuristic algorithm for continuous engineering optimization: harmony search theory and practice. Computer Methods in Applied Mechanics and Engineering 194, 3902–3933 (2005)
5. Yang, X.-S.: Nature-inspired metaheuristic algorithms. Luniver, Frome (2008)
6. Ingram, G., Zhang, T.: An introduction to the harmony search algorithm. In: Rangaiah, G.P. (ed.) Stochastic global optimization: techniques and applications in chemical engineering. World Scientific, Singapore (accepted)
7. Geem, Z.W., Kim, J.H., Loganathan, G.V.: A new heuristic optimization algorithm: harmony search. Simulation 76, 60–68 (2001)
8. Lee, K.S., Geem, Z.W.: A new structural optimization method based on the harmony search algorithm. Computers & Structures 82, 781–798 (2004)
9. Lee, K.S., Geem, Z.W., Lee, S.-H., et al.: The harmony search heuristic algorithm for discrete structural optimization. Engineering Optimization 37, 663–684 (2005)
10. Geem, Z.W.: Optimal cost design of water distribution networks using harmony search. Engineering Optimization 38, 259–277 (2006)
11. Li, L., Chi, S.-C., Lin, G.: Genetic algorithm incorporated with harmony procedure and its application to searching of non-circular critical slip surface in soil slopes. Shuili Xuebao / Journal of Hydraulic Engineering 36, 913–918 (2005) (in Chinese)

12. Geem, Z.W., Hwangbo, H.: Application of harmony search to multi-objective optimization for satellite heat pipe design. In: Proceedings of US-Korea Conference on Science, Technology, and Entrepreneurship (UKC 2006), Teaneck, NJ, USA, (CD-ROM), August 10–13 (2006)
13. Lee, K.S., Choi, C.S.: Discrete-continuous configuration optimization methods for structures using the harmony search algorithm. Key Engineering Materials 324-325, 1293–1296 (2006)
14. Ayvaz, M.T.: Simultaneous determination of aquifer parameters and zone structures with fuzzy c-means clustering and meta-heuristic harmony search algorithm. Advances in Water Resources 30, 2326–2338 (2007)
15. Mahdavi, M., Chehreghani, M.H., Abolhassani, H., et al.: Novel meta-heuristic algorithms for clustering web documents. Applied Mathematics and Computation 201, 441–451 (2008)
16. Forsati, R., Haghighat, A.T., Mahdavi, M.: Harmony search based algorithms for bandwidth-delay-constrained least-cost multicast routing. Computer Communications 31, 2505–2519 (2008)
17. Geem, Z.W., Williams, J.C.: Harmony search and ecological optimization. International Journal of Energy and Environment 1, 150–154 (2007)
18. Mahdavi, M., Fesanghary, M., Damangir, E.: An improved harmony search algorithm for solving optimization problems. Applied Mathematics and Computation 188, 1567–1579 (2007)
19. Cheng, Y.M., Li, L., Chi, S.C.: Performance studies on six heuristic global optimization methods in the location of critical slip surface. Computers and Geotechnics 34, 462–484 (2007)
20. Geem, Z.W.: Novel derivative of harmony search algorithm for discrete design variables. Applied Mathematics and Computation 199, 223–230 (2008)
21. Mukhopadhyay, A., Roy, A., Das, S., et al.: Population-variance and explorative power of harmony search: an analysis. In: Second National Conference on Mathematical Techniques: Emerging Paradigms for Electronics and IT Industries (MATEIT 2008), New Delhi, India, September 26–28 (2008)
22. Geem, Z.W., Kim, T.G., Kim, J.H.: Optimal layout of pipe networks using harmony search. In: 4th International Conference on Hydro-Science and -Engineering, Seoul, Korea, (CD-ROM), September 26–29 (2000)
23. Geem, Z.W., Park, Y.: Harmony search for layout of rectilinear branched networks. WSEAS Transactions on Systems 5, 1349–1354 (2006)
24. Geem, Z.W.: Harmony search algorithm for the optimal design of large-scale water distribution network. In: 7th International IWA Symposium on Systems Analysis and Integrated Assessment in Water Management (Watermatex 2007), Washington DC, USA, (CD-ROM) May 7–9 (2007)
25. Al-Betar, M.A., Khader, A.T., Gani, T.A.: A harmony search algorithm for university course timetabling. In: 7th International Conference on the Practice and Theory of Automated Timetabling (PATAT 2008), Montreal, Canada, August 18–22 (2008)
26. Li, H.-Q., Li, L.: A novel hybrid particle swarm optimization algorithm combined with harmony search for high dimensional optimization problems. In: 2007 International Conference on Intelligent Pervasive Computing (IPC 2007), Jeju Island, Korea, October 11–13 (2007)
27. Omran, M.G.H., Mahdavi, M.: Global-best harmony search. Applied Mathematics and Computation 198, 643–656 (2008)

28. Li, L., Chi, S.-C., Zheng, R.-M., et al.: Mixed search algorithm of non-circular critical slip surface of soil slop. China Journal of Highway and Transport 20, 1–6 (2007) (in Chinese)
29. Lo, K.-C.: A HS-DLM hybrid searching algorithm for structural optimization. Masters thesis, National Central University, Taiwan (2008) (in Chinese)
30. Geem, Z.W., Kim, J.H., Loganathan, G.V.: Harmony search optimization: application to pipe network design. International Journal of Modelling and Simulation 22, 125–133 (2002)
31. Geem, Z.W.: Comparison harmony search with other meta-heuristics in water distribution network design. In: 8th Annual Water Distribution Systems Analysis Symposium (WDSA 2006), Cincinnati, USA, (CD-ROM), August 27–30 (2006)
32. Geem, Z.W., Tseng, C.-L.: Engineering applications of harmony search. In: Late-Breaking Papers of Genetic and Evolutionary Computation Conference (GECCO 2002), New York City, USA, July 9–13 (2002)
33. Geem, Z.W.: Improved harmony search from ensemble of music players. In: Gabrys, B., Howlett, R.J., Jain, L.C. (eds.) KES 2006. LNCS (LNAI), vol. 4251, pp. 86–93. Springer, Heidelberg (2006)
34. Yu, L., Xin, K.: Harmony search optimization for design of water distribution systems. Sciencepaper Online (2007) (accessed November 6, 2007), http://www.paper.edu.cn/en/paper.php?serial_number=200705-157
35. Paik, K.-R., Jeong, J.-H., Kim, J.-H.: Use of a harmony search for optimal design of coffer dam drainage pipes. Journal of Korean Society of Civil Engineers 21, 119–128 (2001) (in Korean)
36. Geem, Z.W.: Harmony search optimisation to the pump-included water distribution network design. Civil Engineering and Environmental Systems (2008), doi:10.1080/10286600801919813
37. Kim, S., Lee, H.: The leak detection method using GA and HS algorithm. In: World Water & Environmental Resources Congress 2003, Philadelphia, USA, June 23–26 (2003)
38. Kim, S.-H., Yoo, W.-S., Oh, K.-J., et al.: Transient analysis and leakage detection algorithm using GA and HS algorithm for a pipeline system. Journal of Mechanical Science and Technology 20, 426–434 (2006)
39. Kim, J.-H., Baek, C.-W., Jo, D.-J., et al.: Optimal planning model for rehabilitation of water networks. Water Science and Technology: Water Supply 4, 133–147 (2004)
40. Baek, C.W., Kim, E.S., Park, M.J., et al.: Development of optimal decision-making system for rehabilitation of water distribution systems using ReHS. Journal of Korea Water Resources Association 38, 199–212 (2005) (in Korean)
41. Geem, Z.W.: Harmony search in water pump switching problem. In: Wang, L., Chen, K., S. Ong, Y. (eds.) ICNC 2005. LNCS, vol. 3612, pp. 751–760. Springer, Heidelberg (2005)
42. Geem, Z.W., Tseng, C.-L.: New methodology, harmony search and its robustness. In: Late-Breaking Papers of Genetic and Evolutionary Computation Conference (GECCO 2002), New York City, USA, July 9–13 (2002)
43. Geem, Z.W.: Optimal scheduling of multiple dam system using harmony search algorithm. In: Sandoval, F., Prieto, A.G., Cabestany, J., Graña, M. (eds.) IWANN 2007. LNCS, vol. 4507, pp. 316–323. Springer, Heidelberg (2007)
44. Kim, J.H., Geem, Z.W., Kim, E.S.: Parameter estimation of the nonlinear Muskingum model using harmony search. Journal of the American Water Resources Association 37, 1131–1138 (2001)
45. Paik, K., Kim, J.H., Kim, H.S., et al.: A conceptual rainfall-runoff model considering seasonal variation. Hydrological Processes 19, 3837–3850 (2005)

46. Saka, M.P.: Optimum geometry design of geodesic domes using harmony search algorithm. Advances in Structural Engineering 10, 595–606 (2007)
47. Erdal, F., Saka, M.P.: Effect of beam spacing in the harmony search based optimum design of grillages. Asian Journal of Civil Engineering (Building and Housing) 9, 215–228 (2008)
48. Degertekin, S.: Optimum design of steel frames using harmony search algorithm. Structural and Multidisciplinary Optimization 36, 393–401 (2008)
49. Degertekin, S.O.: Harmony search algorithm for optimum design of steel frame structures: a comparative study with other optimization methods. Structural Engineering and Mechanics 29, 391–410 (2008)
50. Geem, Z.W., Lee, K.S., Tseng, C.-L.: Harmony search for structural design. In: Genetic and Evolutionary Computation Conference (GECCO 2005), Washington DC, USA, June 25–29 (2005)
51. Li, L.-J., Huang, Z., Liu, F.: An improved particle swarm optimizer for truss structure optimization. In: Wang, Y., Cheung, Y.-m., Liu, H. (eds.) CIS 2006. LNCS (LNAI), vol. 4456, pp. 1–10. Springer, Heidelberg (2007)
52. Li, L.J., Huang, Z.B., Liu, F., et al.: A heuristic particle swarm optimizer for optimization of pin connected structures. Computers & Structures 85, 340–349 (2007)
53. Liu, F., Huang, Z., Li, L., et al.: Heuristic particle swarm optimizer and its application to structural optimization. Yingyong Jichu yu Gongcheng Kexue Xuebao / Journal of Basic Science and Engineering 16, 57–64 (2008) (in Chinese)
54. Jang, W.S., Kang, H.I., Lee, B.H.: Hybrid simplex-harmony search method for optimization problems. In: 2008 IEEE Congress on Evolutionary Computation (CEC 2008), Hong Kong, China, June 1–6 (2008)
55. Fesanghary, M., Mahdavi, M., Minary-Jolandan, M., et al.: Hybridizing harmony search algorithm with sequential quadratic programming for engineering optimization problems. Computer Methods in Applied Mechanics and Engineering 197, 3080–3091 (2008)
56. Ryu, S., Duggal, A.S., Caspar, N.H., et al.: Mooring cost optimization via harmony search. In: 26th International Conference on Offshore Mechanics and Arctic Engineering (OMAE 2007), San Diego, USA, June 10–15 (2007)
57. Gao, X.Z., Wang, X., Ovaska, S.J.: Modified harmony search methods for uni-modal and multi-modal optimization. In: 8th International Conference on Hybrid Intelligent Systems (HIS 2008), Barcelona, Spain, September 10–12 (2008)
58. Tian, Y.-H., Bo, Y.-M., Gao, M.-F.: Harmony annealing algorithm for multi-dimensional function optimization. Computer Simulation 21, 79–82 (2004) (in Chinese)
59. Tian, Y.-H., Bo, Y.-M., Gao, M.-F.: Parameters choice criteria in harmony annealing for function optimization. Computer Simulation 22, 70–74+89 (2005) (in Chinese)
60. Li, H.-Q., Li, L.: A novel hybrid real-valued genetic algorithm for optimization problems. In: 2007 International Conference on Computational Intelligence and Security (CIS 2007), Harbin, China, December 15–19 (2007)
61. Li, Q., Yang, S., Ruan, Y.: A hybrid algorithm for optimizing multi-modal functions. Wuhan University Journal of Natural Sciences 11, 551–554 (2006)
62. Li, L., Wang, Y.-J., Wang, Q.-S., et al.: New procedure for simulating arbitrary slip surface of soil slope in stability analysis. Shuili Xuebao / Journal of Hydraulic Engineering 39, 535–541 (2008) (in Chinese)
63. Li, L., Chi, S.-C., Chu, X.-S.: Location of non-circular slip surface using the modified harmony search method based on correcting strategy. Yantu Lixue / Rock and Soil Mechanics 27, 1714–1718 (2006) (in Chinese)

64. Li, L., Chi, S.-C., Lin, G., et al.: Slope stability analysis using extremum principle by Pan Jiazheng and harmony search method. Yantu Lixue / Rock and Soil Mechanics 28, 157–162 (2007) (in Chinese)
65. Li, L., Chi, S.-C., Cheng, Y.-M., et al.: Improved genetic algorithm and its application to determination of critical slip surface with arbitrary shape in soil slope. Shuili Xuebao / Journal of Hydraulic Engineering 38, 157–162 (2007) (in Chinese)
66. Li, L., Chi, S., Cheng, Y., et al.: Improved genetic algorithm and its application to determination of critical slip surface with arbitrary shape in soil slope. Frontiers of Architecture and Civil Engineering in China 2, 145–150 (2008)
67. Li, L., Chi, S., Lin, G.: Chaos harmony search method and its application to local factor of safety method for soil slopes. Yanshilixue Yu Gongcheng Xuebao / Chinese Journal of Rock Mechanics and Engineering 25, 2763–2769 (2006) (in Chinese)
68. Li, L., Chen, Z.-Y., Chi, S.-C., et al.: Three-dimensional slope stability analysis based on NURBS simulation. Yantu Gongcheng Xuebao / Chinese Journal of Geotechnical Engineering 30, 212–218 (2008) (in Chinese)
69. Li, L., Chi, S.-C.: Application of new version of harmony search algorithm in slope stability analysis. Journal of Water Resources and Architectural Engineering 5, 1–6+24 (2007) (in Chinese)
70. Cheng, Y.M., Li, L., Lansivaara, T., et al.: An improved harmony search minimization algorithm using different slip surface generation methods for slope stability analysis. Engineering Optimization 40, 95–115 (2008)
71. Forsati, R., Mahdavi, M., Kangavari, M., et al.: Web page clustering using harmony search optimization. In: 21st Canadian Conference on Electrical and Computer Engineering (CCECE 2008), Niagara Falls, Canada, May 4–7 (2008)
72. Wang, X., Gao, X.Z., Ovaska, S.J.: A hybrid optimization method for fuzzy classification systems. In: 8th International Conference on Hybrid Intelligent Systems (HIS 2008), Barcelona, Spain, September 10–12 (2008)
73. Malaki, M., Pourbaghery, J.A., Abolhassani, H.: A combinatory approach to fuzzy clustering with harmony search and its applications to space shuttle data. In: SCIS & ISIS 2008, Nagoya, Japan, September 17–21 (2008)
74. Forsati, R., Mahdavi, M., Haghighat, A.T., et al.: An efficient algorithm for bandwidth-delay constrained least cost multicast routing. In: 21st Canadian Conference on Electrical and Computer Engineering (CCECE 2008), Niagara Falls, Canada, May 4–7 (2008)
75. Cruz, F., Coelho, A., Reis, L.P.: Automatic parameterization for expeditious modelling of virtual urban environments: a new hybrid meta heuristic. In: 5th International Conference on Informatics in Control, Automation and Robotics (ICINCO 2008), Funchal, Portugal, May 11–15 (2008)
76. Geem, Z.W., Lee, K.S., Park, Y.: Application of harmony search to vehicle routing. American Journal of Applied Sciences 2, 1552–1557 (2005)
77. Geem, Z.W.: School bus routing using harmony search. In: Genetic and Evolutionary Computation Conference (GECCO 2005), Washington DC, USA, June 25–29 (2005)
78. Geem, Z.W., Tseng, C.-L., Park, Y.-J.: Harmony search for generalized orienteering problem: Best touring in china. In: Wang, L., Chen, K., S. Ong, Y. (eds.) ICNC 2005. LNCS, vol. 3612, pp. 741–750. Springer, Heidelberg (2005)
79. Ceylan, H., Ceylan, H., Haldenbilen, S., et al.: Transport energy modeling with meta-heuristic harmony search algorithm, an application to Turkey. Energy Policy 36, 2527–2535 (2008)

80. Fesanghary, M., Damangir, E., Soleimani, I.: Design optimization of shell and tube heat exchangers using global sensitivity analysis and harmony search algorithm. Applied Thermal Engineering (2008), doi:10.1016/j.applthermaleng.2008.05.018
81. Vasebi, A., Fesanghary, M., Bathaee, S.M.T.: Combined heat and power economic dispatch by harmony search algorithm. International Journal of Electrical Power & Energy Systems 29, 713–719 (2007)
82. Amor, J., Alexandre, E., Gil-Pita, R., et al.: Music-inspired harmony-search algorithm applied to feature selection for sound classification in hearingssss aids. In: Audio Engineering Society 124th Convention, Amsterdam, The Netherlands, May 17–20 (2008)
83. Panchal, A.: Harmony search optimization for HDR prostate brachytherapy. American Association of Physicists in Medicine (2008) (accessed October 29, 2008), `http://chapter.aapm.org/midwest/spring08/panchal.pdf`
84. Dong, H., Bo, Y., Gao, M., et al.: Improved harmony search for detection with photon density wave. In: Proceedings of SPIE, vol. 6625, p. 662523 (2008)
85. Liu, T.-N., Feng, Z.-B.: Adaptive identification and filtering based on harmony search. Journal of Jilin University (Information Science Edition) 22, 306–309 (2004) (in Chinese)
86. Tian, Y.-H., Bo, Y.-M., Gao, M.-F.: The application of harmony annealing algorithm for predictive control of nonlinear systems. Automation in Petro-Chemical Industry 2, 39–42 (2005) (in Chinese)
87. Santos Coelho, L.d., de Andrade Bernert, D.L.: An improved harmony search algorithm for synchronization of discrete-time chaotic systems. Chaos, Solitons & Fractals (2008), doi:10.1016/j.chaos.2008.09.028
88. Geem, Z.W.: Harmony search algorithm for solving sudoku. In: Apolloni, B., Howlett, R.J., Jain, L. (eds.) KES 2007, Part I. LNCS (LNAI), vol. 4692, pp. 371–378. Springer, Heidelberg (2007)
89. Geem, Z.W., Choi, J.-Y.: Music composition using harmony search algorithm. In: Giacobini, M. (ed.) EvoWorkshops 2007. LNCS, vol. 4448, pp. 593–600. Springer, Heidelberg (2007)
90. Zarei, O., Fesanghary, M., Farshi, B., et al.: Optimization of multi-pass face-milling via harmony search algorithm. Journal of Materials Processing Technology (2008), doi:10.1016/j.jmatprotec.2008.05.029

Harmony Search Methods for Multi-modal and Constrained Optimization

X.Z. Gao, X. Wang, and S.J. Ovaska

Department of Electrical Engineering, Helsinki University of Technology,
Otakaari 5 A, FI-02150 Espoo, Finland
{gao,xiaolei}@cc.hut.fi, seppo.ovaska@tkk.fi

Abstract. The Harmony Search (HS) method is an emerging meta-heuristic optimization algorithm. In this chapter, we propose two modified HS methods to handle the multi-modal and constrained optimization problems. The first modified HS method employs a novel HS memory management approach to handle the multi-modal problems. The second modified HS method utilizes the Pareto-dominance technique, and it targets at the constrained problems. Several simulation examples are used to demonstrate and verify the effectiveness of our new HS methods.

Keywords: Harmony Search, Multi-Modal Optimization, Constrained Optimization, Artificial Fish Swarm, Pareto Dominance.

1 Introduction

Firstly proposed by Geem *et al.* in 2001 [1], the HS method is inspired by the underlying principles of the musicians' improvisation of the harmony. During the recent years, it has been successfully applied to the areas of function optimization [2], mechanical structure design [3], pipe network optimization [4] and Sudoku puzzle [5]. However, empirical study has shown that the original HS method has a limitation in dealing with the multi-modal and constrained optimization problems. To overcome these drawbacks, we propose two modified HS methods in this chapter. The first modified HS method employs an efficient diversity maintenance policy for the members of the HS memory. The second modified HS method is based on the direct handling of the given constraints. Extensive simulations have demonstrated that our two modified HS methods can outperform the original HS method in attacking the multi-modal and constrained problems.

The rest of this chapter is organized as follows. We briefly introduce the essential principles of the HS method in Section 2. The two new HS methods are presented and discussed in Sections 3 and 4, respectively. Simulation examples of the multi-modal and constrained optimization are demonstrated in Section 5. Finally, in Section 6, we conclude our chapter with some remarks and conclusions.

2 Harmony Search Method

When musicians improvise a harmony, they usually try various possible combinations of the music pitches stored in their memory. This kind of effective search for a perfect harmony is analogous to the procedure of finding an optimal solution in engineering problems. The HS method is inspired by the working principles of the harmony improvisation [1]. Figure 1 shows the flowchart of the basic HS method, in which there are four principal steps involved.

Z.W. Geem (Ed.): Music-Inspired Harmony Search Algorithm, SCI 191, pp. 39–51.
springerlink.com

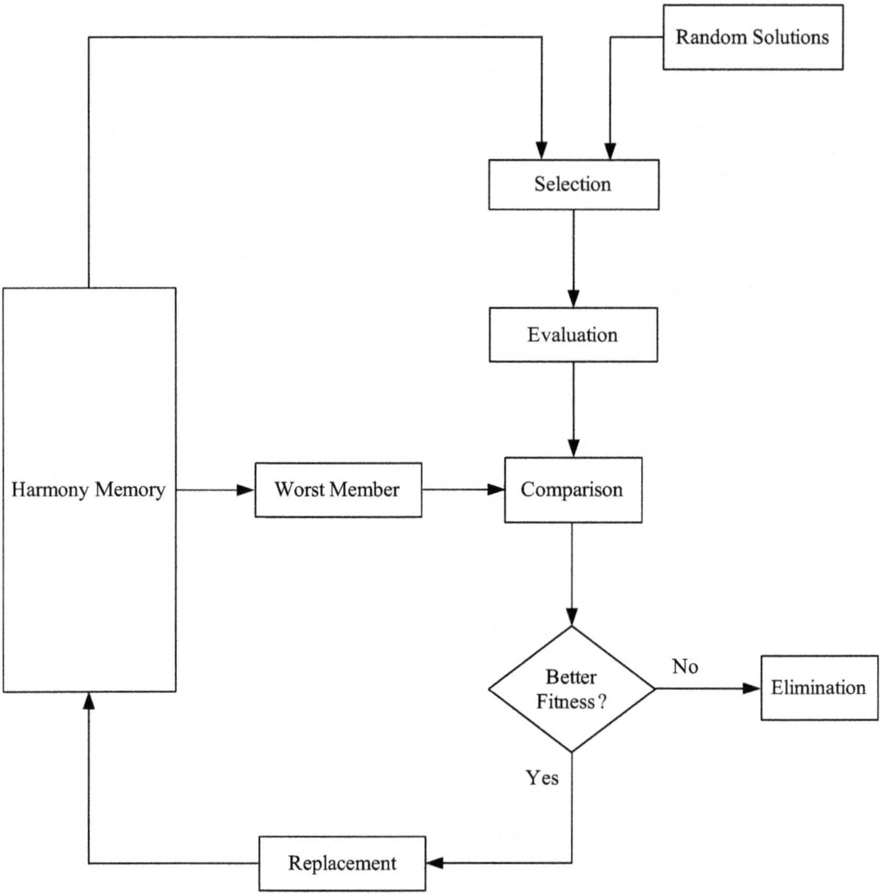

Fig. 1. Harmony Search (HS) method

Step 1. Initialize a harmony memory (HM). The initial HM consists of a certain number of randomly generated solutions for the optimization problem under consideration. For an n-dimension problem, an HM with the size of HMS can be represented as follows:

$$\mathrm{HM} = \begin{bmatrix} x_1^1, x_2^1, \cdots, x_n^1 \\ x_1^2, x_2^2, \cdots, x_n^2 \\ \vdots \\ x_1^{HMS}, x_2^{HMS}, \cdots, x_n^{HMS} \end{bmatrix}, \tag{1}$$

where $\left(x_1^i, x_2^i, \cdots, x_n^i\right)$ ($i = 1, 2, \cdots, HMS$) is a candidate solution. HMS is typically set to be between 10 and 100.

Step 2. Improvise a new solution $\left(x_1', x_2', \cdots, x_n'\right)$ from the HM. Each component of this solution, x_j', is obtained based on the Harmony Memory Considering Rate

(HMCR). The HMCR is defined as the probability of selecting a component from the HM members, and 1-HMCR is, therefore, the probability of generating it randomly. If x'_j comes from the HM, it can be further mutated according to the Pitching Adjust Rate (PAR). The PAR determines the probability of a candidate from the HM to be mutated. Obviously, the improvisation in HS is rather similar to the production of off-spring in genetic algorithm (GA) [6] with the mutation and crossover operations. However, GA creates new chromosomes using only one (mutation) or two (simple crossover) existing ones, while the generation of new solutions in the HS method makes full use of all the HM members.

Step 3. Update the HM. First, the new solution from Step 2 is evaluated. If it yields a better fitness than that of the worst member in the HM, it will replace that one. Otherwise, it is eliminated.

Step 4. Repeat Step 2 to Step 3 until a termination criterion (e.g., maximal number of iterations) is met.

Similar to the GA and particle swarm algorithms [7], the HS method is a random search technique. It does not require any prior domain knowledge, such as gradient information of the objective function. However, different from those population-based approaches, it only utilizes a single search memory to evolve. Therefore, the HS method has the distinguished feature of algorithm simplicity. Note that the HS memory stores the past search experiences, and plays an important role in its optimization performance. In the next section, we employ a fish swarm-based strategy to improve the management of the HM members so that the HS method can cope with the multi-modal optimization problems.

3 Modified HS Method for Multi-modal Optimization

Multi-modal optimization is an important but challenging topic in the field of optimization [8]. Unfortunately, it is difficult for the HS method to locate all the global optima of the multi-modal problems, because the HM members can be easily stagnated into one or several of them during iteration. Thus, the key issue of applying the HS method to the multi-modal optimization is to effectively maintain the diversity of the HM members. The regular HM management policy has been explained in Section 2. However, some additional approaches are needed to determine whether a solution from Step 2 can replace the worst member in the HM. Indeed, the qualification of a candidate solution as a new HM member should be based on not only its fitness but also its similarity to the existing members.

Inspired by the artificial fish swarm algorithm [9], we propose a new control mechanism, as shown in Fig. 2, for updating the HM in our modified HS method so as to attack the multi-modal problems. Suppose the fitness of the current HM members is denoted as f_i ($i = 1,2,\cdots,HMS$). After a new candidate solution $(x'_1, x'_2, \cdots, x'_n)$, with the fitness f', is obtained, we first measure its distances d_i ($i = 1,2,\cdots,HMS$) to all HM members:

$$d_i = \left\| [x'_1, x'_2, \cdots, x'_n] - [x^i_1, x^i_2, \cdots, x^i_n] \right\|, \tag{2}$$

where $\|\ \|$ is an appropriately selected distance metric.

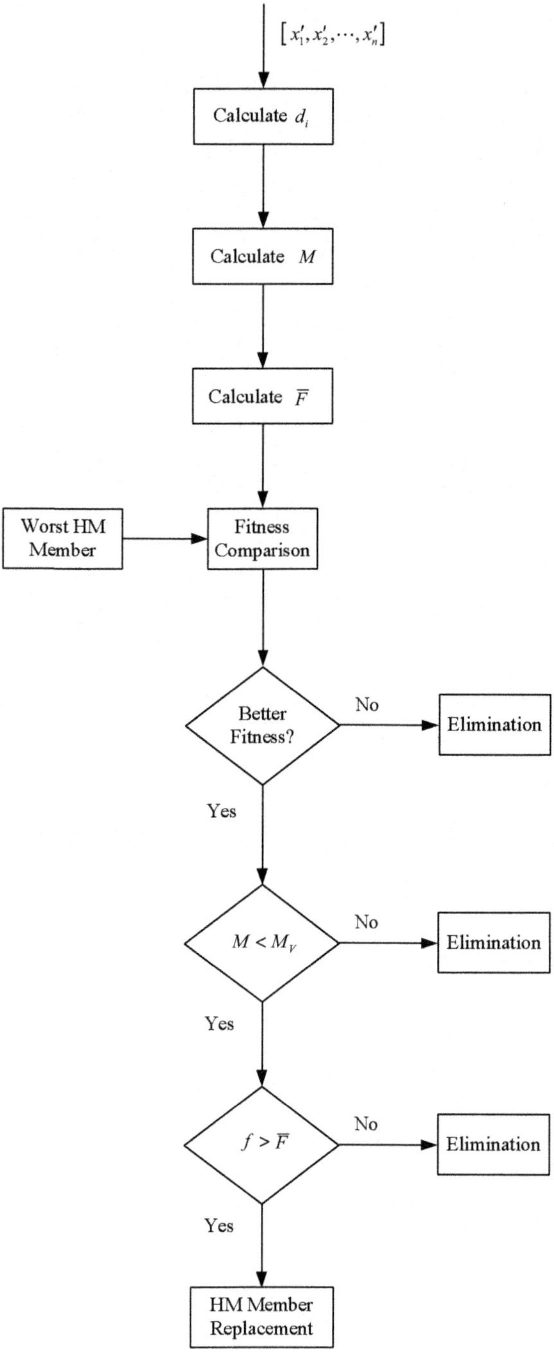

Fig. 2. HM member control in modified HS method for multi-modal optimization

Next, we calculate the number of the HM members, M , which are in the vicinity V of $\left(x_1', x_2', \cdots, x_n'\right)$. In other words, only the HM members, whose d_i are smaller than V , are counted here. The average fitness of these HM members, $\overline{F}$, is given as follows:

$$\overline{F} = \frac{\sum_{i=1}^{M} f_i}{M}. \tag{3}$$

Thus, $\left(x_1', x_2', \cdots, x_n'\right)$ will replace the worst member of the HM only if it meets the following three conditions:

1) f' is better than that of the worst HM member,
2) M is smaller than a preset threshold M_V ,
3) f' is better than $\overline{F}$.

It is concluded from the above explanations that our approach can prevent the over-similarity among the HM members so that the diversity of the HS solutions can be maintained. That is to say, the modified HS method is well-suited for handling the multi-modal problems. Nevertheless, the proposed technique has two drawbacks. Firstly, parameters V and M_V are always applications dependent, and are usually chosen based on *trial and error*. They can considerably affect the multi-modal optimization performance of the modified HS method. Unfortunately, there is no analytic way yet to guarantee their best values. Secondly, as in (2), the distances between $\left(x_1', x_2', \cdots, x_n'\right)$ and all the present HM members have to be calculated. This requirement can certainly result in a time-consuming procedure, in case of a large HMS .

4 Modified HS Method for Constrained Optimization

4.1 Constrained Optimization Problems

Most of the practical optimization problems are actually constrained optimization problems, whose goal is to find an optimal solution that satisfies certain given constraints [10]. In general, a constrained optimization problem is described as follows:

$$\text{find } \vec{x} = (x_1, x_2, \cdots, x_n) \text{ to minimize } \mathrm{f}(\vec{x}),$$

$$\text{subject to } \mathrm{g}_i(\vec{x}) \le 0,\ i = 1,2,\cdots,I \text{ and } \mathrm{h}_j(\vec{x}) = 0,\ j = 1,2,\cdots,J$$

where $\mathrm{f}(\vec{x})$ is the objective function, and $\mathrm{g}_i(\vec{x})$ and $\mathrm{h}_j(\vec{x})$ are the inequality and equality constraint functions, respectively. As a matter of fact, the equality constraint functions can be transformed into inequality constraint functions:

$$\left|\mathrm{h}_j(\vec{x})\right| - \varepsilon \le 0, \tag{4}$$

where ε is a small enough tolerance parameter. Therefore, we here only consider the inequality constraint functions $g_i(\vec{x}) \leq 0$, $i = 1,2,\cdots,I$.

The constrained optimization problems are generally difficult to deal with, because the constraint functions can divide the whole search space into disjoint islands. Numerous constraint-handling techniques have been proposed and investigated during the past decades [11]. One popular solution is to define a new fitness function $F(\vec{x})$ to be optimized [11]. $F(\vec{x})$ is the combination of the objective function $f(\vec{x})$ and weighted penalty terms $P_i(\vec{x})$, $i = 1,2,\cdots,I$, which reflect the violation of the constraint functions:

$$F(\vec{x}) = f(\vec{x}) + \sum_{i=1}^{I} w_i P_i(\vec{x}), \tag{5}$$

where w_i ($i = 1,2,\cdots,I$) are the preset weights. The overall optimization performance depends on the penalty terms and their weights, and may significantly deteriorate with inappropriately chosen ones. In this section, we propose a modified HS method for the direct handling of these constraints.

4.2 Modified HS Method for Constrained Optimization

As aforementioned, the new HM members are generated either from the existing HM members or in a random way. Nevertheless, they are not guaranteed to always meet all the constraints. Figure 3 shows that the new HM members, which satisfy all the constraints, can be acquired based upon *trial and error*. This task is exhaustive, especially for complex constraint functions.

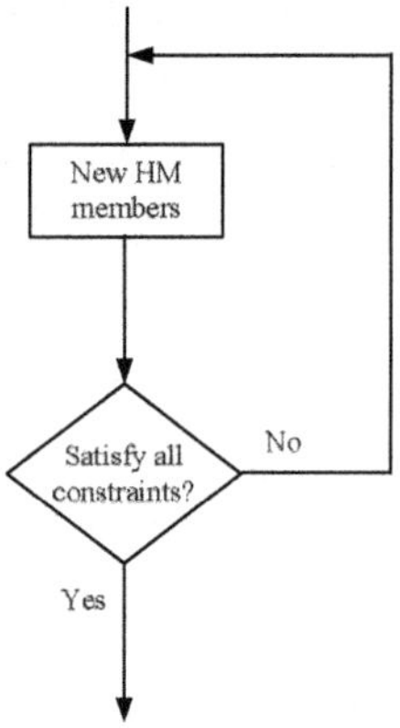

Fig. 3. Generation of new HM members by trial and error method

In our modified HS method, we make full use of those HM members that do not even meet the constraints. The key issue is how to rank the HM members according to their objective as well as constraint function values. Here, the values of the constraint functions of the HM members are stored together with their objective function values in the HM. The HM members are divided into two parts: feasible members and infeasible members. The former satisfy all the constraint functions while the latter do not. Thus, the ranking of the HM members consists of two corresponding stages: ranking of the feasible HM members and ranking of the infeasible ones. The ranking of the feasible HM members is straightforward: they can be sorted using their objective function values. However, for the infeasible ones, the ranking is based on the Pareto-dominance of the HM members [12]. An infeasible HM member dominates another member if none of its constraint function values is larger and at least one is smaller. Formally, the Pareto-dominance is defined as follows. Suppose there are two infeasible HM members, $\vec{x}^1$ and $\vec{x}^2$. If

$$\forall i \in \{1,2,\cdots,I\}: g_i(\vec{x}^1) \le g_i(\vec{x}^2) \wedge \exists i \in \{1,2,\cdots,I\}: g_i(\vec{x}^1) < g_i(\vec{x}^2),$$

we conclude that $\vec{x}^1$ dominates $\vec{x}^2$. For each infeasible HM member, we can calculate the number of others that dominate it, which implies its relative degree of violation of the constraint functions. That is, the rank of an infeasible HM member is determined by the number of other infeasible HM members by which it is dominated.

After the HM is ranked, the worst HM member $\vec{x}^{\#}$ can be selected and compared with a new solution candidate $\vec{x}^*$. Note, $\vec{x}^*$ does not need to be feasible. When $\vec{x}^{\#}$ is compared with $\vec{x}^*$, $\vec{x}^*$ will replace $\vec{x}^{\#}$ only in one of the following three cases:

1) $\vec{x}^*$ is feasible, and $\vec{x}^{\#}$ is infeasible
2) both $\vec{x}^*$ and $\vec{x}^{\#}$ are feasible, and $f(\vec{x}^*) < f(\vec{x}^{\#})$
3) both $\vec{x}^*$ and $\vec{x}^{\#}$ are infeasible, and $\vec{x}^*$ dominates $\vec{x}^{\#}$

More precisely, $\vec{x}^*$ replaces $\vec{x}^{\#}$ on condition that

$$\begin{cases} \forall i \in \{1,2,\cdots,I\}: g_i(\vec{x}^*) \le 0 \\ \wedge \\ \exists i \in \{1,2,\cdots,I\}: g_i(\vec{x}^{\#}) > 0 \end{cases} \vee \begin{cases} \forall i \in \{1,2,\cdots,I\}: g_i(\vec{x}^*) \le 0 \\ \wedge \\ \forall i \in \{1,2,\cdots,I\}: g_i(\vec{x}^{\#}) \le 0 \\ \wedge \\ f(\vec{x}^*) < f(\vec{x}^{\#}) \end{cases}$$
$$\vee \begin{cases} \exists i \in \{1,2,\cdots,I\}: g_i(\vec{x}^*) > 0 \\ \wedge \\ \exists i \in \{1,2,\cdots,I\}: g_i(\vec{x}^{\#}) > 0 \\ \wedge \\ \forall i \in \{1,2,\cdots,I\}: g_i(\vec{x}^*) \le g_i(\vec{x}^{\#}) \\ \wedge \\ \exists i \in \{1,2,\cdots,I\}: g_i(\vec{x}^*) < g_i(\vec{x}^{\#}) \end{cases} \quad (6)$$

Figure 4 illustrates this procedure of comparison between $\vec{x}^{\#}$ and $\vec{x}^*$ and replacement of $\vec{x}^{\#}$ with $\vec{x}^*$.

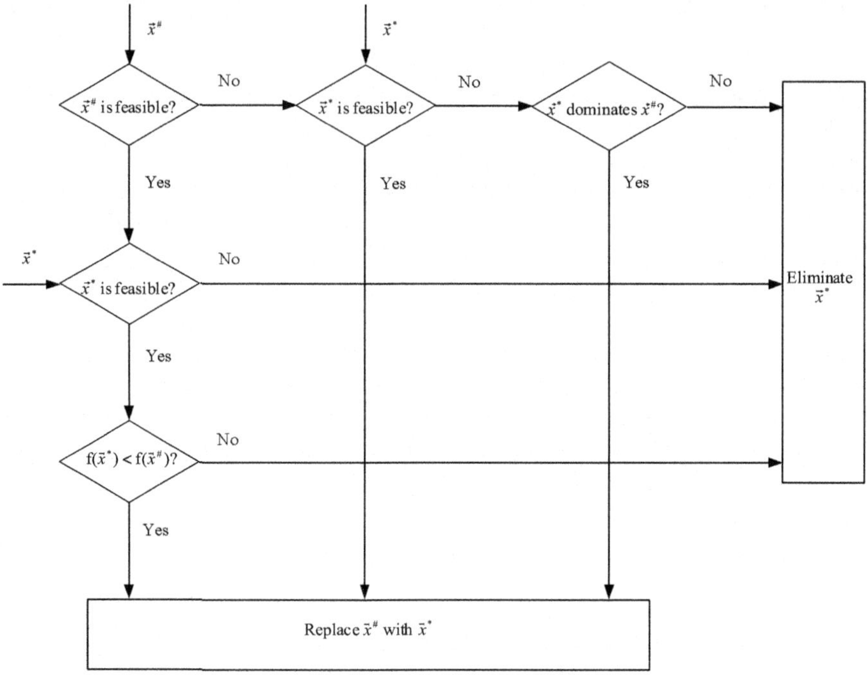

Fig. 4. Comparison between $\vec{x}^{\#}$ and $\vec{x}^{*}$ and replacement of $\vec{x}^{\#}$ with $\vec{x}^{*}$

It has been observed that the infeasible HM members were able to be evolved in the modified HS method. In other words, we do not have to search for new feasible HM members by repeatedly examining them with the constraint functions, as in Fig. 3. Compared with the original HS method, our approach needs only a considerably smaller number of constraint functions evaluation, and thus, can provide more efficient solutions. This advantage will be demonstrated using computer simulations in Section 5.

5 Simulations

In this section, we investigate the effectiveness of the two modified HS methods with simulation examples of the multi-modal and constrained optimization problems.

5.1 Multi-modal Optimization Problems

In this example, the multi-modal optimization capability of our first modified HS method is examined using the following two functions [11]:

$$f_1(x, y) = 200 - (x^2 + y - 11)^2 + (x + y^2 - 7)^2, \quad -5 \le x, y \le 5 \tag{7}$$

$$f_2(x, y) = x\sin(4\pi x) - y\sin(4\pi y + \pi) + 1, \quad -1 \le x, y \le 1 \tag{8}$$

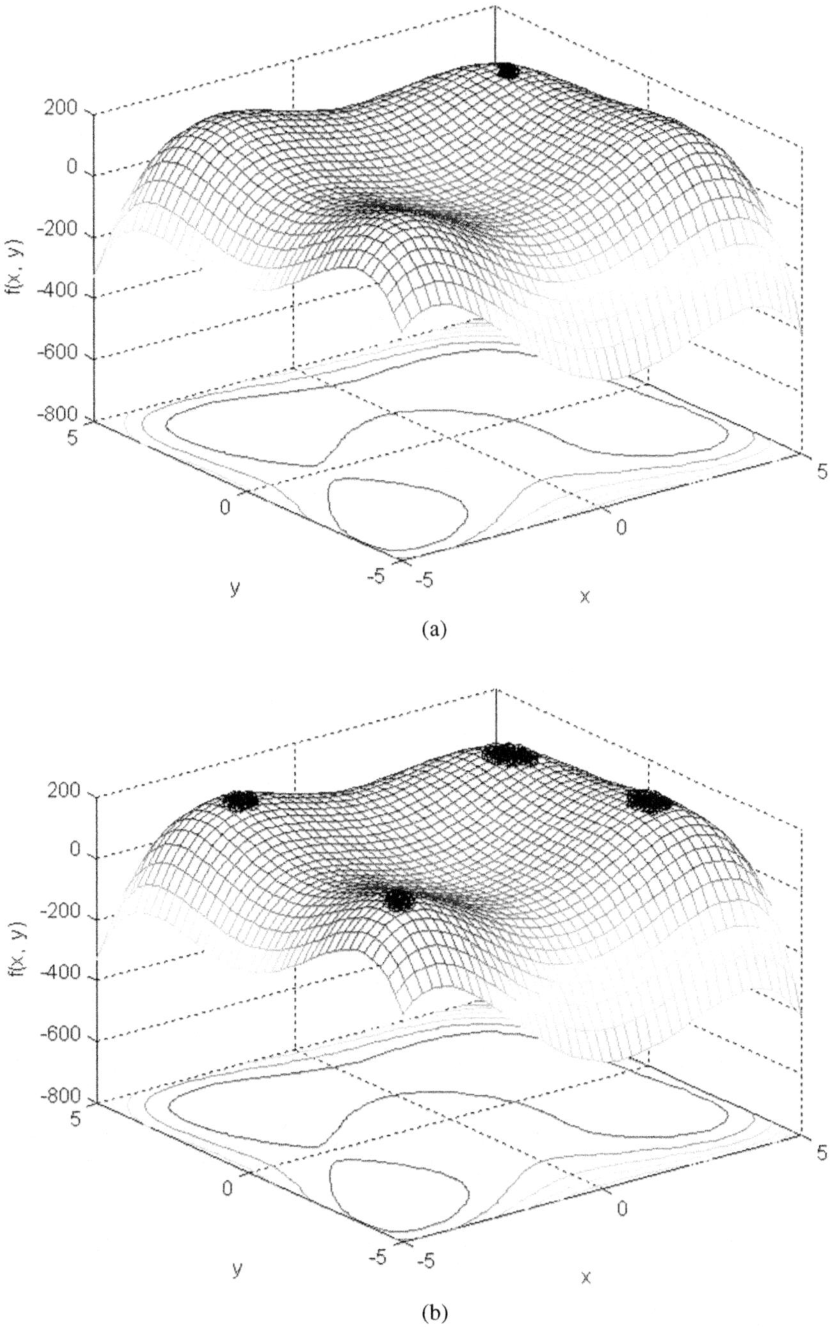

Fig. 5. Optimization results of $f_1(x, y)$ using (a) regular HS and (b) modified HS methods

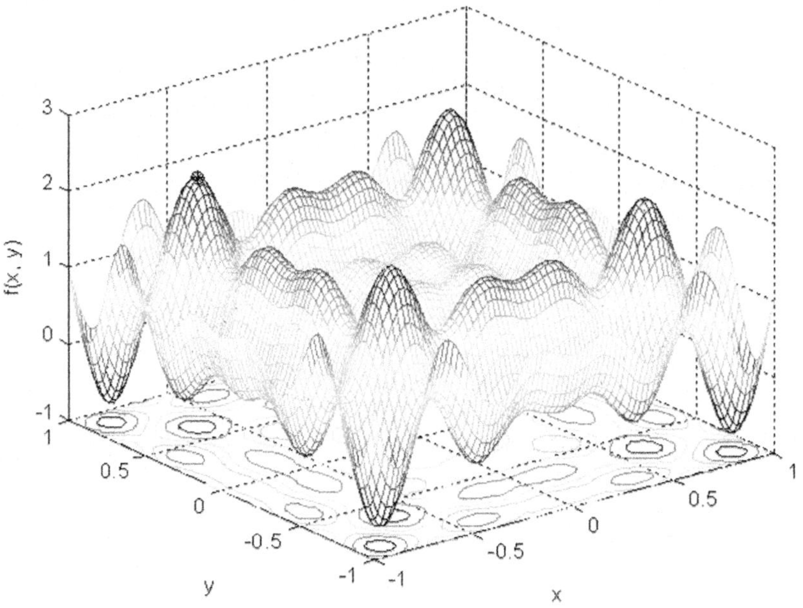

(a)

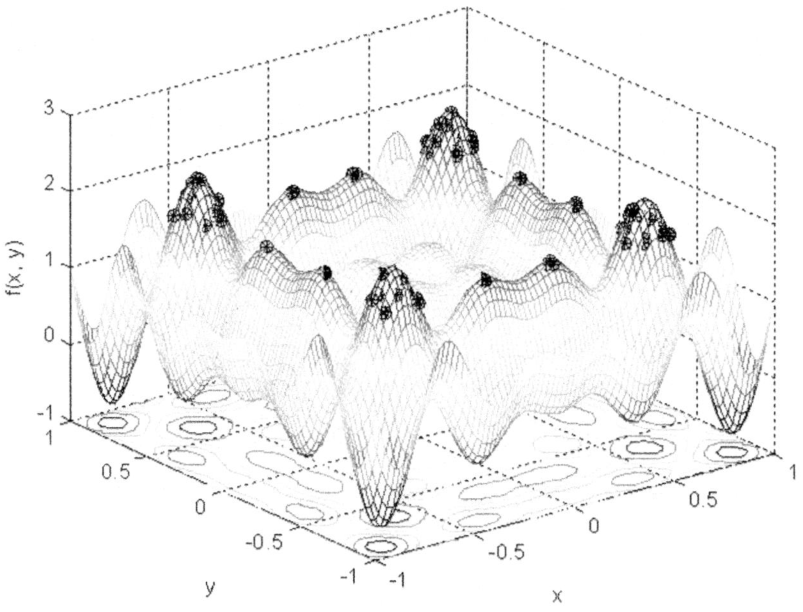

(b)

Fig. 6. Optimization results of $f_2(x, y)$ using (a) regular HS and (b) modified HS methods

Each nonlinear function has only one global optimum (minimum) but several local optima. The goal of the optimization algorithms is to find not only the global optimum but also as many local optima as possible. The optimization results of the regular HS and modified HS methods are illustrated in Figs. 5 and 6. HMS = 100, HMCR = 0.75, and PAR = 0.6 are used in both HS techniques. In our modified HS method, V and M_V for $\mathrm{f}_1(x, y)$ and $\mathrm{f}_2(x, y)$ are:

$$\text{for } \mathrm{f}_1(x, y)\text{, } V = 0.15\text{, and } M_V = 2$$

$$\text{for } \mathrm{f}_2(x, y)\text{, } V = 0.075\text{, and } M_V = 3$$

It is clearly visible that the regular HS method can only find the global optimum, while the modified HS method is capable of locating most of the local optima in addition to the global optimum. However, we emphasize that V and M_V can affect the multi-modal optimization performance of our modified HS method. In case of a fixed M_V, if V is too small, the behaviors of the normal HS and modified HS methods are similar. On the other hand, the performance of the modified HS method can significantly deteriorate in case of a too large V. Unfortunately, like other HS parameters, how to choose the best V and M_V is still an unsolved problem, although some adaptation strategies can be the potential solutions.

5.2 Constrained Optimization Problems

Two numerical examples are used here to demonstrate the capability of the second modified HS method in handling constrained optimization problems. The first example is a function optimization problem with two variables, x_1 and x_2, and two constraints, $\mathrm{g}_1(\vec{x})$ and $\mathrm{g}_2(\vec{x})$, as follows:

$$\text{Minimize } \mathrm{f}(\vec{x}) = (x_1^2 + x_2 - 11)^2 + (x_1 + x_2^2 - 7)^2,$$

subject to

$$\mathrm{g}_1(\vec{x}) = (x_1 - 0.05)^2 + (x_2 - 2.5)^2 - 4.84 \le 0,$$

$$\mathrm{g}_2(\vec{x}) = 4.84 - x_1^2 - (x_2 - 2.5)^2 \le 0,$$

$$0 \le x_1 \le 6,\ 0 \le x_2 \le 6.$$

The modified HS method successfully found the optimal solution without violating any constraint while a gradient-based method (BFGS) failed to find feasible solution.

The second example is the optimal design of the welded beam. The goal here is to minimize the fabricating cost of the welded beam subject to the constraints on the shear stress τ, bending stress on the beam σ, buckling load on the bar P_C, end deflection of the beam δ, and side constraints. The four design variables (h, l, t, and b) are denoted as x_1, x_2, x_3, and x_4, as shown in Fig. 7. The details of the welded beam design problem are:

$$\text{Minimize } f(\vec{x}) = 1.10471 x_1^2 x_2 + 0.04811 x_3 x_4 (x_2 + 14),$$

subject to

$$g_1(\vec{x}) = \tau(\vec{x}) - \tau_{max} \le 0,\ g_2(\vec{x}) = \sigma(\vec{x}) - \sigma_{max} \le 0,$$

$$g_3(\vec{x}) = x_1 - x_4 \le 0,\ g_4(\vec{x}) = \delta(\vec{x}) - \delta_{max} \le 0,\ g_5(\vec{x}) = P - P_C(\vec{x}) \le 0,$$

where

$$\tau(\vec{x}) = \sqrt{(\tau')^2 + 2\tau'\tau''\frac{x_2}{2R} + (\tau'')^2},\ \tau' = \frac{P}{\sqrt{2}x_1x_2},\ \tau'' = \frac{MR}{J},\ M = P(L + \frac{x_2}{2}),$$

$$R = \sqrt{\frac{x_2^2}{4} + (\frac{x_1 + x_3}{2})^2},\ J = 2\left\{\sqrt{2}x_1x_2\left[\frac{x_2^2}{12} + (\frac{x_1 + x_3}{2})^2\right]\right\},\ \sigma(\vec{x}) = \frac{6PL}{x_3^2x_4},$$

$$\delta(\vec{x}) = \frac{4PL^3}{Ex_3^3x_4},\ P_C(\vec{x}) = \frac{4.013E\sqrt{\frac{x_3^2x_4^6}{36}}}{L^2}(1 - \frac{x_3}{2L}\sqrt{\frac{E}{4G}}),$$

$$P = 6000 \text{ lb},\ L = 14 \text{ in},\ \delta_{max} = 0.25 \text{ in},\ E = 30\times10^6 \text{ psi},$$

$$G = 12\times10^6 \text{ psi},\ \tau_{max} = 13{,}600 \text{ psi},\ \sigma_{max} = 30{,}000 \text{ psi},$$

$$0.125 \le x_1 \le 5,\ 0.1 \le x_2, x_3 \le 10,\ 0.1 \le x_4 \le 5.$$

Again, our modified HS method found good solutions without violating any constraint for this practical design problem.

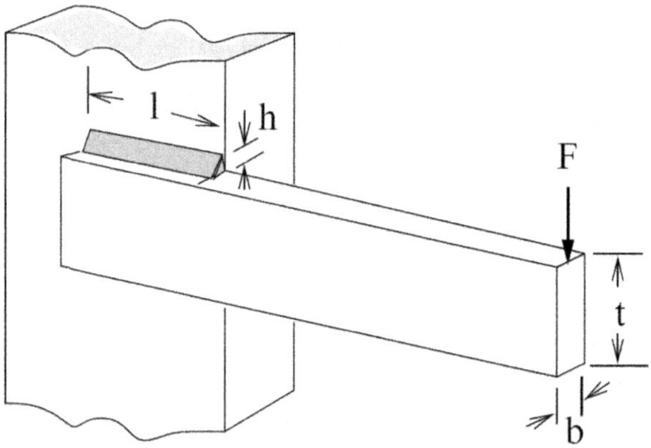

Fig. 7. Schematic of welded beam

6 Conclusions

In this chapter, we proposed and discussed two modified HS methods to deal with the multi-modal and constrained optimization problems. In the first modified HS method, a fish swarm-based technique is employed to maintain the diversity of the HM members, which makes it a suitable candidate for the multi-modal problems. The second modified HS method is capable of directly handling the constraints in the constrained optimization problems. Several simulation examples have been employed to verify the effectiveness of the proposed schemes. Compared with the original HS method, better optimization results are acquired using our modified HS approaches. However, some important theoretical issues, such as convergence analysis, need to be further explored. We are also going to study how to apply these new methods in manipulating more real-world problems.

Acknowledgments

This research work was funded by the Academy of Finland under Grant 214144.

References

1. Geem, Z.W., Kim, J.H., Loganathan, G.V.: A new heuristic optimization algorithm: harmony search. Simulation 76, 60–68 (2001)
2. Lee, K.S., Geem, Z.W.: A new meta-heuristic algorithm for continuous engineering optimization: harmony search theory and practice. Computer Methods in Applied Mechanics and Engineering 194, 3902–3922 (2005)
3. Lee, K.S., Geem, Z.W.: A new structural optimization method based on the harmony search algorithm. Computers and Structures 82, 781–798 (2004)
4. Geem, Z.W., Kim, J.H., Loganathan, G.V.: Harmony search optimization: application to pipe network design. International Journal of Modeling and Simulation 22, 125–133 (2002)
5. Geem, Z.W.: Harmony search algorithm for solving sudoku. In: Apolloni, B., Howlett, R.J., Jain, L. (eds.) KES 2007, Part I. LNCS (LNAI), vol. 4692, pp. 371–378. Springer, Heidelberg (2007)
6. Poli, R., Langdon, W.B.: Foundations of Genetic Programming. Springer, Berlin (2002)
7. Engelbrecht, A.P.: Fundamentals of Computational Swarm Intelligence. John Wiley & Sons Ltd., West Sussex (2005)
8. Goldberg, D.E.: Genetic Algorithms in Search, Optimization, and Machine Learning. Addison-Wesley, Reading (1989)
9. Wang, C.R., Zhou, C.L., Ma, J.W.: An improved artificial fish-swarm algorithm and its application in feed-forward neural networks. In: Proceedings of 2005 International Conference on Machine Learning and Cybernetics, pp. 2890–2894 (2005)
10. Arora, J.S.: Introduction to Optimum Design. McGraw-Hill, New York (1989)
11. Michalewicz, Z.: Genetic Algorithms + Data Structures = Evolution Programs, 3rd edn. Springer, Berlin (1996)
12. Coello, C.A.C.: Constraint-handling in genetic algorithms through the use of dominance-based tournament selection. Advanced Engineering Informatics 16, 193–203 (2002)

Solving NP-Complete Problems by Harmony Search

Mehrdad Mahdavi

Department of Computer Engineering, Sharif University of Technology, Tehran, Iran
mahdavi@ce.sharif.edu

Abstract. In the last few years, there has been explosive growth in the application of Harmony Search (HS) in solving NP-complete problems in computer science. The success of the HS algorithm in finding relatively good solutions to these problems discriminates it as an affirmative alternative to other conventional optimization techniques. This chapter surveys the existing literature on the application of HS in combinatorial optimization problems. We begin by presenting HS based algorithms for solving problems such as Sudoku puzzle, music composition, orienteering problem, and vehicle routing. We then turn to solve a multicast routing problem with two constraints (i.e. bandwidth and delay constraints). Finally, we show how to apply HS to a clustering problem.

Keywords: Harmony Search, Sudoku Puzzle, Music Composition, Orienteering Problem, Vehicle Routing, Multicast Routing, Clustering.

1 Introduction

Many real world applications in computer science are NP-complete problems which have a growing importance in the scientific and industrial world. No polynomial-time algorithm has yet been discovered for NP-complete problems, nor has anyone yet been able to prove that no polynomial-time algorithm can exit for any one of them. In many real life settings, high quality solutions to these problems such as multicast routing, clustering, and vehicle routing are required in a very short amount of time, and many algorithms to tackle them have been developed. In cases, especially when large scale problems are considered, metaheuristics are one of the best alternatives people can often rely on since exact algorithms take exponential time to find an optimal solution. This chapter deals with some harmony search applications in computer science, especially NP-complete problems, for example: a Sudoku puzzle, a music composition, a generalized orienting problem, a vehicle routing problem, a multicast routing problem, and a document clustering problem.

The Sudoku puzzle consists of a partially completed row-column grid of cells partitioned into blocks, to be filled with a prescribed set of N distinct symbols (typically the numbers $\{1, ..., N\}$), so that each row, column and block contains exactly one of each element of the set. The puzzle can be solved using a variety of algorithms. This chapter shows how efficiently and effectively Sudoku puzzle can be solved using the HS algorithm.

Music composition is analogous to optimization in some ways and could be formulated as an optimization problem which is solvable by metaheuristic algorithms. Also, HS mimics musicians' behaviors such as random play, memory-based play, and pitch-adjusted play when they perform improvisation. Thus, there clearly exists similarity. This chapter shows the HS application with music composition.

Z.W. Geem (Ed.): Music-Inspired Harmony Search Algorithm, SCI 191, pp. 53–70.
springerlink.com

The HS was also applied to a TSP-like NP-hard generalized orienteering problem (GOP) which is to find the utmost route under the total distance limit while maximizing multiple goals. Here, GOP is a variant of an orienteering problem (OP) which is used to model many practical problems.

Another problem which is considered in this chapter is a school bus routing problem which is a multi-objective problem with a goal to minimize both the number of operating buses and the travel time of all buses, with two major constraints (bus capacity and time window). This kind of problem has many applications such as inventory routing, customer or vehicle assignment, production scheduling, and postal delivery or school bus routing. The HS algorithm could find global optimum within far less function evaluations than total enumeration.

The advent of various real-time multimedia applications in high-speed networks creates a need for quality of service (QoS) based multicast routing. Two important QoS constraints are bandwidth constraint and end-to-end delay constraint. The QoS based multicast routing problem is a known NP-complete problem that depends on (1) bounded end-to-end delay and link bandwidth along the paths from the source to each destination, and (2) minimum cost of the multicast tree. This chapter considers the application of HS in multicast routing to generate multicast trees with minimum cost satisfying the bandwidth and delay constraint.

Finally, HS is applied to document clustering which has a crucial role in organizing information and search engine results, enhancing web crawling, and information retrieval or filtering. Since a clustering problem has NP-complete nature, the larger the size of the problem, the harder it is to find the optimal solution and the longer it is to reach a reasonable result. So it is reasonable to model a clustering problem as an optimization and apply metaheuristic algorithms such as HS.

2 Sudoku Puzzle

Sudoku, which is Japanese term meaning 'one-digit number,' is the new craze in logic puzzles. The Sudoku puzzle consists of 9×9 grids and 3×3 blocks for a total of 81 cells. Each puzzle, which has a unique solution, has some cells that already have given-numbers. The objective of the puzzle is to fill in the remaining cells with the numbers 1 through 9 so that every row, column, and 3×3 submatrix has a sum total of 45.

2.1 Problem Formulation for Sudoku Puzzles

The Sudoku puzzle could be formulated as a binary integer programming for general $n \times n$ puzzles. Let x_{ijk} equal one if element (i, j) of the $n \times n$ matrix of Sudoku contains the integer k, and otherwise equal to zero. The formulation is as follows:

$$\text{Minimize } \mathbf{0}^{\mathrm{T}}\mathbf{x} \tag{1}$$

s.t.

$$\sum_{i=1}^{n} x_{ijk} = 1, \quad 1 \le j, k \le n \tag{2}$$

$$\sum_{j=1}^{n} x_{ijk} = 1, \quad 1 \le i, k \le n \tag{3}$$

$$\sum_{i,j \in G_l} x_{ijk} = 1, \quad 1 \le k, l \le n \tag{4}$$

$$\sum_{k=1}^{n} x_{ijk} = 1, \quad 1 \le i, j \le n \tag{5}$$

$$\sum_{k=1}^{n} x_{ijk} = 1, \quad \forall (i, j) \in N \tag{6}$$

$$x_{ijk} \in \{0,1\} \tag{7}$$

where G_l stands for the submatrix l, and N contains elements in the matrix, which have no value in the initial puzzle. Constraints 2, 3, and 4 guarantee that only one k exists in each column, row, and submatrix respectively. Constraint 5 forces every position in the matrix to be filled. Please note that it is possible to formulate the problem as an integer linear program (ILP) by letting the integer variables take the values from 1 to 9.

In [1], Geem employed a formulation with penalties as follows:

$$\text{Minimize } Z = \sum_{i=1}^{9} \left| \sum_{j=1}^{9} x_{ij} - 45 \right| + \sum_{j=1}^{9} \left| \sum_{i=1}^{9} x_{ij} - 45 \right| + \sum_{l=1}^{9} \left| \sum_{(i,j) \in G_l} x_{ij} - 45 \right| \tag{8}$$

where x_{ij} = cell at row i and column j, which has an integer value from 1 to 9.

The first term in Equation 8 represents the penalty function for each horizontal row; the second term for each vertical column; and the third term for each block. It should be noted that, although the sum of each row, each column, or each block equals 45, it does not guarantee that the numbers 1 through 9 are used exactly once. However, any violation of the uniqueness affects other row, column, or block which contains the wrong value jointly.

2.2 Example Puzzle

The HS algorithm application was applied to the Sudoku puzzle proposed by Nicolau and Ryan [2] with 40 given values (white cells in Figure 1). The total searching space for this case is $9^{41} = 1.33 \times 10^{39}$ if integer programming formulation is considered.

The proposed HS model found the unique global optimum without any row, column or block violation after 285 function evaluations, taking 9 seconds on Intel Celeron 1.8 GHz Processor as shown in Figure 1 [1].

	1	2	3	4	5	6	7	8	9
1	2	5	4	3	1	6	8	9	7
2	7	6	3	9	8	5	1	2	4
3	1	9	8	4	2	7	6	5	3
4	9	8	1	7	5	3	2	4	6
5	6	3	2	8	4	9	7	1	5
6	5	4	7	2	6	1	9	3	8
7	4	7	5	6	9	2	3	8	1
8	3	1	9	5	7	8	4	6	2
9	8	2	6	1	3	4	5	7	9

Fig. 1. Final Solution of Sudoku Puzzle Obtained by HS

3 Music Composition

Music can be composed by metaheuristic algorithms. Horner and Goldberg [3] applied a GA model to bridge-music composition in the minimalist style. The fitness function in their model was the degree of pattern match and duration. Ralley [4] proposed another GA model to develop music melody. However, the melody developed by GA could not be evaluated because there was no appropriate fitness function. Biles [5] developed an interactive GA model, GenJam, to play jazz solos. The GenJam model has been applied to many jazz tunes. This section shows another music composition example using HS.

3.1 Formulation of Medieval Music Composition

Gregorian chant is a monophonic unaccompanied song of the Roman Catholic Church in the middle age, and organum is an early form of polyphonic music accompanying the Gregorian chant.

The organum has the following composing characteristics: 1) the original harmony line progresses in parallel since it starts on the same note; 2) for the parallel motion, the interval of perfect fourth is frequently used while those of perfect fifth and unison (or octave) are also preferred; 3) in order to distinguish the vox principalis (chant melody) from vox organalis (harmony), the former should always be located above the latter. The above-mentioned organum composition technique can be formulated as the following optimization problem:

$$\text{Minimize} \sum_{i=1}^{N} Rank(x_i) + \sum_{i=1}^{N} Penalty(x_i),\ i = 1, 2, \cdots, N \tag{9}$$

s.t.

$$x_i \le m_i,\ i=1,2,\cdots,N \tag{10}$$

$$| x_i - x_{i-1} | \le | m_i - m_{i-1} |,\ i=2,3,\cdots,N \tag{11}$$

$$x_{Start} = m_{Start} \tag{12}$$

$$x_{End} = m_{End} \tag{13}$$

$$x_i \in \{\text{Do}, \text{Re}, \text{Mi}, \text{Fa}, \text{Sol}, \text{La}, \text{Si}, \text{Do}^+\} \tag{14}$$

where x_i is the ith pitch in harmony line and m_i is the ith pitch in original chant line. Equation 9 represents the fitness function, which is the summation of the rank term and the penalty term for each pitch in the harmony line (vox organalis). As the interval of a perfect fourth between vox principalis and vox organalis is most preferred, it receives the highest priority (rank = 1). The ranking for intervals is a prior knowledge and explicitly determined. Equation 10 represents the constraint that organum pitch should be lower than or equal to corresponding chant pitch. Equation 11 represents the constraint that the interval between two consecutive organum pitches should be less than or equal to that in two consecutive chant pitches. Equations 12 and 13 represent boundary conditions that constrain starting and ending pitches in organum.

3.2 Example Composition

Figure 2 shows a Gregorian chant 'Rex caeli Domine' as well as a corresponding organum composed by HS [6]. The upper line in the figure is the Gregorian chant melody and the lower line is the organum line. There are 28 pitches (number of decision variables) accompanied by the organum line, which represents $8^{28} (= 1.93 \times 10^{25})$ combinatorial possibilities.

Fig. 2. Organum Composed by the HS algorithm

The HS model composed the organum line by generating up to 3,000 improvisations within one second on Intel Celeron 1.8GHz CPU and obtained aesthetically pleasing organum as shown in Figure 2. Also, a more complex organum piece, which has 50 pitches, was also successfully composed using the same process.

4 Generalized Orienteering Problem

The orienteering problem is a subset selection version of well-known traveling salesman problem (TSP) with profits. The objective of the OP is to construct a path starting at an origin and ending at a destination which maximizes the total profit without violating a prescribed travel distance limit. Due to the fact that distance is limited, tours may not include all locations. It should be noted that the OP is equivalent to the TSP when the distance is relaxed just enough to cover all points and the start and end points are not specified. The OP is used to model many practical problems such as inventory routing, customer or vehicle assignment, and production scheduling. Additionally, a variation of OP with time windows has applications to problems including postal delivery or school bus routing.

The GOP, proposed by Wang et al. [7], is a generalization of the orienteering problem. The main difference between the two is that each city in GOP has multiple scores while each city in OP has only one score. So, the objective of GOP is to find the optimal tour under the constraint of total distance limit while satisfying multiple goals rather than only one goal as is the case with OP [8].

4.1 Problem Formulation for OP and GOP

The objective of the OP problem is to find an acyclic path from a start point to an end point so that the total score collected from the visited points will be maximized without violating the given distance constraint (i.e., the total length of the edges forming the path should be less than a specified limit). The OP problem is a known NP-hard problem.

The GOP differs from OP because each point in GOP has more than one score. The score vector is expressed as $\mathbf{S}(i) = (S_1(i), S_2(i), \ldots, S_m(i))^T$, where m is the number of individual goals. A differentiable objective function, which defines the total score of path P, can be formulated as follows:

$$Z = \sum_{g=1}^{m} W_g \left[\left\{ \sum_{i \in P} [S_g(i)]^k \right\}^{1/k} \right] \tag{15}$$

where W_g is the weight of goal g, and the exponent k is a controlling constant which is set to 5 in the example tackled in the following section.

4.2 Eastern China Touring Example

The GOP example using eastern China, which has 27 cities and 4 goals, was first proposed in [7]. If a traveler visits the eastern part of China, as shown in Figure 3, and the person wants to travel to as many cities as possible with the purpose of best fulfilling multiple goals such as maximizing: 1) natural beauty, 2) historical interest, 3) cultural event, and 4) business opportunities under the limited total moving distance, the travel can become a generalized orienteering problem where each city has certain quantified scores for all goals. The tour is evaluated based on the summation of those scores obtained from all locations forming the tour.

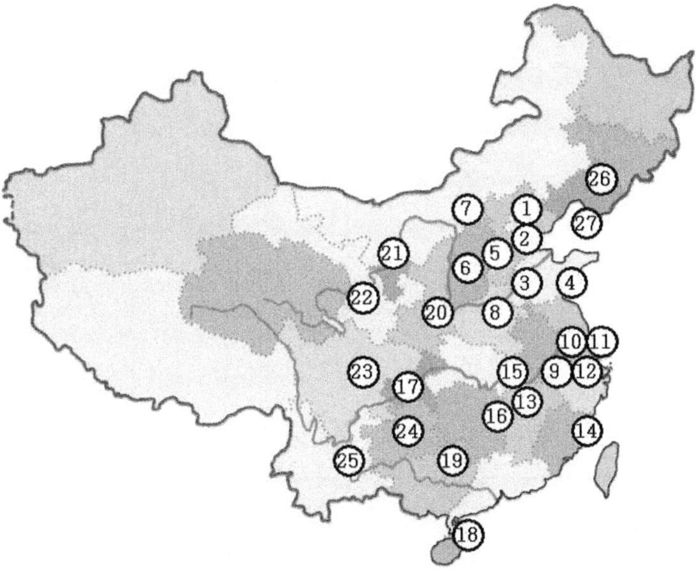

Fig. 3. Map of 27 cities in eastern part of China

The HS model [9] was developed to address this problem and compared with the artificial neural network model [7]. Here, the distance limit equals to 5,000 km. Compared to the results of ANN, HS could find better solutions three times out of five cases.

5 Vehicle Routing Problem

The vehicle routing problem (VRP) is a combinatorial optimization problem seeking to pick up, deliver or simply give a service to customers dispersed in an area with a fleet of vehicles. Each vehicle has a certain capacity and each customer has a certain demand. Further, there exist a depot(s) and a distance (length, cost, time) matrix among customers. In VRP, we look for optimal vehicle routes (minimum distance or number of vehicles). The school bus routing problem (SBRP) is a special case of VRP. From a school's perspective, the SBRP aims to provide students with an efficient and equitable transportation service.

An example of a network to be optimized consisting of one bus depot, one school, and ten bus stops is shown in Figure 4. Each bus stop is demanded by certain number of commuting students, and travel time (in minutes) between two stops is specified on the link. The objective of the problem is to minimize both the total number of operating buses and the moving time of the buses.

The HS algorithm could find the global optimum ($307,980) only after 1,000 improvisations (or function evaluations), which was examined by the total enumeration for this $4^{10} (\approx 1.05 \times 10^6)$ combinatorial problem [10]. It took 6.6 seconds on Intel 233MHz CPU to search 0.1% of total solution space $(= 10^3 / 1.05 \times 10^6)$.

HS results were also examined by comparing them with those of GA. In order to fairly compare HS with GA, the number of objective function evaluations and the

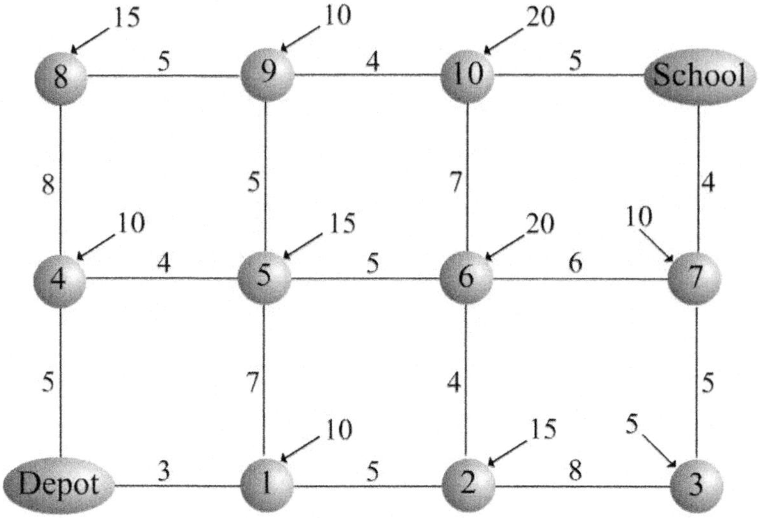

Fig. 4. Schematic of school bus routing network

number of computational runs are the same in both algorithms. In HS, the number of function evaluations is 1,000 and the number of runs is 20 with different HMS, HMCR, and PAR. In GA, the number of function evaluations is 1,000 (= population size × number of generations) and the number of runs is 20 with different population size, crossover rate, and mutation rate, recommended by Koumousis and Georgiou [11].

Both algorithms could find the global optimum solution. However, HS reached it twice while GA did it once out of 20 different runs. The average costs were $399,870 in HS, and $409,597 in GA. In addition, HS performed each run slightly faster than GA on the same machine.

6 QoS Based Multicast Routing

Many newer applications on the Internet fit into the model where the same information is sent to multiple receivers with varying quality of service (QoS) constraints concurrently, such as an audio conference, interactive multimedia games and web-base learning. These applications are better supported by a model termed multicast where information is sent to a specific group simultaneously and is only duplicated for delivery purposes when necessary in the most efficient way. Two important QoS constraints are the bandwidth constraint and the end-to-end delay constraint. The QoS based multicast routing problem is a known NP-complete problem that depends on (1) a bounded end-to-end delay and link bandwidth along the paths from the source to each destination, and (2) a minimum cost of the multicast tree.

Accordingly, multicast routing plays a critical role in the bulk of these new applications. Designing multicast routing algorithms is technically challenging and a complicated task to deliver multimedia information in a timely, smooth, synchronized manner over a decentralized, shared network environment.

Generally, there are two approaches to solve this problem: (1) obtaining an optimal solution in exponential time; (2) obtaining near-optimal solutions by metaheuristic algorithms. Though the first pursues an optimal solution, it is sometimes impractical due to its NP-hard nature. Thus the second approach can be a feasible alternative instead. Some metaheuristics for the delay-constrained or more constraint multicast routing problems have been studied in the literature [12-15]. The simulation results given by Salama et al. [16] have shown that most of the metaheuristic algorithms either work too slowly or cannot compute the delay constrained multicast tree with least cost.

6.1 Problem Formulation of QoS Based Multicast Routing

Communication networks consist of nodes connected via links. The nodes are the originators and receivers of information, while the links serve as the transport between nodes. A network is modeled as a directed weighted graph *G=(V, E)* where *V* is a finite set of nodes representing routers or switches and *E* is a set of edges representing communication links between network nodes. Three non-negative real-valued functions are associated with each link e(e∈ *E*): cost *C*(e): $E\rightarrow R^+$, delay *D*(e): $E\rightarrow R^+$, and available bandwidth *B*(e): $E\rightarrow R^+$. Function *D*(e) represents the delay that the packet experiences on a link including queuing, transmission, and propagation delay and *B*(e) is bandwidth where *C*(e) defines the link cost function. *C*(e) defines the measure we want to optimize (minimize), *D*(e)and *B*(e) define the criteria we want to constrain (bound).

A multicast tree $T(s,D)$ is a tree *routed* at *s* and spanning all members of $D \subseteq V-\{s\}$. We refer to *D* as the *destination group* and $s \cup D$ as the *multicast group*. Furthermore, the multicast tree may contain relay nodes, called *Steiner nodes*. The Steiner node is a node that belongs to the multicast tree but not to the multicast group. Let $P_T(s,d)$ be a unique path in the tree *T* from the source node *s* to a destination node $d \in D$. The cost of the tree $T(s,D)$ is the sum of the costs of all links in that tree and is as follow:

$$Cost(T)=\sum_{e\in T(s,D)} C(e) \tag{16}$$

The total path delay from *s* to any destination *d*, is simply the sum of the delay of edges along $P_T(s,d)$, i.e.

$$Delay(P_T(s,d))=\sum_{e\in P_T(s,d)} D(e) \tag{17}$$

The bottleneck bandwidth of the path from *s* to any destination *d*, denoted by Bandwidth($P_T(s,d)$) is defined as the minimum available residual bandwidth at any link along the path:

$$Bandwidth(P_T(s,d))=\min\{B(e), e\in P_T(s,d)\} \tag{18}$$

For notational convenience, we use Δ_d to represent delay constraint and β_d to represent the bandwidth constraint of the destination node *d* in a multicast tree *T*. Based on these definitions, the bandwidth-delay constrained least-cost multicast problem is defined as minimization of Cost (*T*) subject to:

$$\begin{cases} Delay(P_T(s,d)) \le \Delta_d & ,\forall d\in D \\ Bandwidth(P_T(s,d)) \ge \beta_d & ,\forall d\in D \end{cases} \tag{19}$$

It has been demonstrated that finding a feasible multicast tree with two independent additive path constraints is NP-complete. Obviously, since the delay and cost are independent, a bandwidth-delay constrained least-cost multicast routing is a problem of NP-complete.

Let $F(T(s,D))$ be a function that return fitness of a multicast tree $T(s,D)$. $F(T(s,D))$ has the property that, if the $T(s,D)$ is a bandwidth-delay-constrained tree, then, it must return the cost of a tree as its fitness value. Also, $F(T(s,D))$ must be able to estimate how close is an infeasible solution, which does not satisfy constraints, is from the feasible region of search space by handling constraints. The most common approach to handle constraints is to use penalties in $F(T(s,D))$.

According to these properties, we propose the following relation to evaluate solutions:

$$F(T(s,D)) = \frac{\alpha}{\sum_{e \in T(s,D)} C(e)} \prod_{d \in D} \Phi(Delay(P(s,d) - \Delta_d)) \times \prod_{d \in D} \Psi(Bandwidth(P(s,d) - \beta_d)) \quad (20)$$

where α is a positive real factor, $\Phi(\cdot)$ and $\Psi(\cdot)$ are penalty functions. For an additional illustration, an example is shown in Figure 5. The solution obtained using the HS algorithm with $\Delta = 25$ is shown in bold lines in Figure 6.

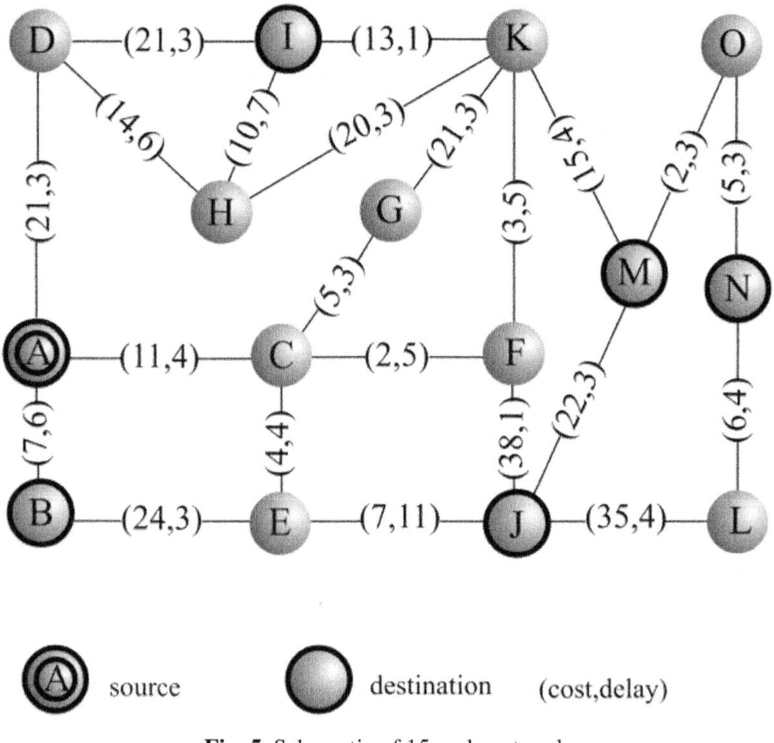

Fig. 5. Schematic of 15 node network

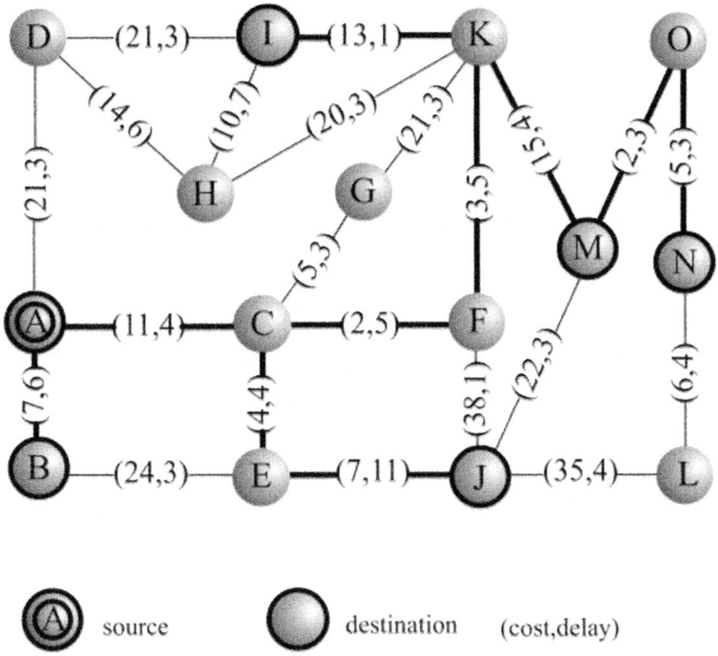

Fig. 6. Solution by HS for 15 node network

One of the most important decisions in applying evolutionary algorithms to the multicast problem is how to encode the solutions. Prüfer numbers offer a deceptively elegant coding of spanning trees. In a complete undirected graph with n nodes, there are n^{n-2} distinct spanning trees.

In edge set representation, each tree is directly represented as a set of its edges. In characteristic vector representation, each solution is represented by a binary vector of |E| bits in which kth bit indicates whether kth edge is or is not part of the tree. Predecessor coding is another representation which designates a root for the tree and then records the predecessor of each node in the tree. In NPI each tree is traversed with breadth-first search (BFS) or depth-first search (DFS) algorithms and each node with its parent's index is inserted to a list in consecutive cells. After determining the representation method, the solutions are encoded in the target representation and inserted into the solutions' pool.

6.2 Application on Real Benchmark Networks

In recent years various optimization methodologies have been developed for constructing optimal multicast constrained trees. SA [17], GA [18-23] and TS [24-29] are general high-level procedures that coordinate simple heuristics and rules to find good approximate solutions to computationally difficult combinatorial optimization problems. These methods have been previously employed to solve the problem of multicast routing, and results showed that these methods are suitable to achieve near-optimal results within a reasonable time [22, 23, 26, 30]. Two algorithms are proposed in [31] for solving the mentioned problem by HS. The first algorithm, named HSPR, is based on Prüfer representation and the second one, called HSNPI, is based on Node Parent Representation (NPI) encoding proposed in [31, 32].

HSPR, HSNPI, and the GA-based algorithm [23] are applied to real data networks. These real networks are obtained from three synthetic instances of the OR-Library called steinb10, steinb11, and steinb18 which are used for the graphical Steiner tree problem. The networks were modified for multicast routing in [31].

Table 1 summarizes the results obtained by applying the HS and GA-based algorithms on the instance networks from the OR-Library. The statistically significant best solutions have been shown in bold. The results provide evidence that the HS outperforms GA in all except for three instances. Table 1 shows that the ratio of the high-quality solutions for HS ranges from 93% to 98%. However, considering both the tractability of an NP-hard problem and the low time complexity of the problem, it is still of practical sense. Table 1 shows that for the small groups the cost deviation of HS is approximately zero and solutions are very close to the optimal one.

Table 1. Simulation results for HS and GA on real topologies

Network Topology	# of Multicast Group	Optimal Cost	**HS**		**GA**	
			Mean # of Iterations	Success Ratio (%)	Mean # of Iterations	Success Ratio (%)
Steinb10	10	75	18,891	**97**	31,215	98
Steinb10	20	162	20,026	93	32,024	**92**
Steinb10	30	234	17,142	**95**	28,023	93
Steinb11	10	62	16,043	**98**	24,888	96
Steinb11	20	145	16,803	**99**	26,902	98
Steinb11	30	237	27,651	96	24,539	**96**
Steinb15	20	174	19,446	**98**	26,052	97
Steinb15	30	261	24,871	**97**	34,558	94
Steinb15	40	328	20,623	94	27,340	**95**

The analysis presented in [31] on real and random networks shows that while the problem of optimal constrained multicast routing is NP-hard [33], the meta-heuristic model based on HS produces good solutions. Also empirical performance shows that the HSNPI algorithm results in the best overall performance with regard to total tree cost in a comparison with a recently developed algorithm based on GA and a modified version of heuristic BSMA algorithm.

Figure 7 shows the execution time of HSPR, HSNPI, BSMA [13], and GA [23] algorithms on a randomly generated network with a different size. As seen in Figure 7, the HSNPI and BSMA models yield very competitive execution time, especially when network nodes are smaller than 45. When they are smaller than 45, the execution time of HSNPI model is almost same as BSMA model. Although the BSMA has a slightly better execution time than HSNPI, the difference in average cost of HSNPI

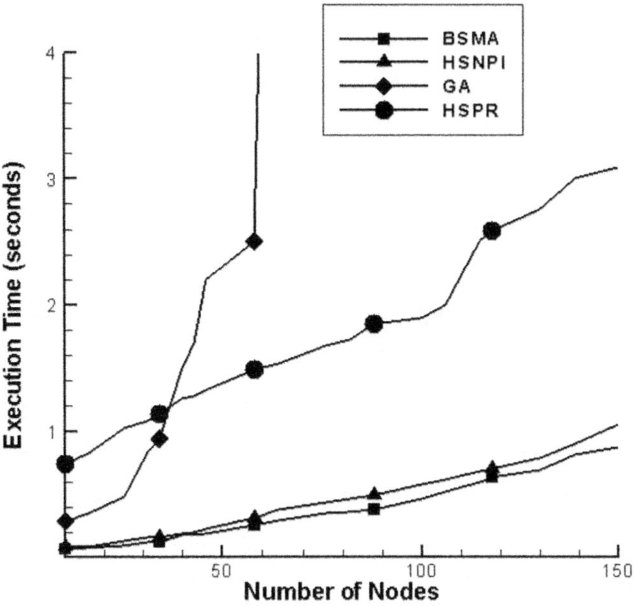

Fig. 7. Execution time of different algorithms on randomly generated networks

in comparison with other algorithms is tremendous and HSNPI outperforms GA when the cost of generated multicast trees is considered.

7 Document Clustering

Fast and high quality document clustering has become an increasingly important technique for enhancing search engine results, web crawling, unsupervised document organization, and information retrieval or filtering. Clustering involves dividing a set of documents into a specified number of groups. The documents within each group should exhibit a large degree of similarity and the similarity among different clusters should be minimized. Some of the more familiar clustering methods are: partitioning algorithms based on dividing entire data into dissimilar groups, hierarchical methods, density and grid based clustering, and some graph based methods [34, 35].

In most document clustering algorithms, documents are represented using a vector-space model. In this model, each document d is considered to be a vector $\vec{d} = \{d_1, d_2, \ldots, d_t\}$ in term-space (set of document 'words') where d_i is the weight of dimension i in vector space and t is the number of term dimensions. The most widely used weighting approach for term weights is the combination of Term Frequency and Inverse Document Frequency (TF-IDF) [36-38].

The similarity between two documents must be measured in some way if a clustering algorithm is to be used. The vector space model gives us a good opportunity for defining different metrics for similarity between two documents. The most common similarity metrics are Minkowski distances [39] and the cosine measure [36, 38, 40].

Minkowski distances computes the distance of documents d and d' by Equation 21 (for n = 2 it is converted to Euclidean distance).

$$D_n(d,d') = \left(\sum_{i=1}^{t} |d_i - d'_i|^n \right)^{1/n} \tag{21}$$

The cosine measure is defined by Equation 22 where $d^T \cdot d'$ is the inner product (dot-product) of two vectors.

$$\cos(d,d') = \frac{d^T \cdot d'}{|d||d'|} \tag{22}$$

Both metrics are widely used in the text document clustering literatures. It seems, however, in the cases when the number of dimensions of two vectors differs largely, the cosine is more useful. In cases where two vectors have almost the same dimension, Minkowski distance can be useful.

As mentioned above, the objective of a clustering algorithm is to maximize the intra-similarity and minimize the inter-similarity. The cohesiveness of clusters can be used as a measure of cluster similarity. The quality of a specific clustering can be determined by Average Distance of Documents to the Cluster Centroid (ADDC) represented by that clustering. This value is measured by the equation:

$$f = \frac{\sum_{i=1}^{k} \left[\frac{\sum_{j=1}^{n_i} D(c_i, d_{ij})}{n_i} \right]}{K} \tag{23}$$

where K is the number of clusters; n_i is the numbers of documents in cluster i (e.g., $n_i = \sum a_{ij}$ $(1 \le j \le n)$); D is distance function, d_{ij} is the jth document of cluster i, and the c_i in the centroid of ith cluster which is the average of all document vectors in the cluster.

Since our goal is to optimize an objective function, clustering is essentially a search problem. Let us stress that dividing n data into K clusters gives rise to a huge number of possible partitions, which is expressed in the form of the Stirling number:

$$\frac{1}{K!} \sum_{i=1}^{K} (-1)^{K-i} \binom{K}{i} i^n \tag{24}$$

This illustrates that the clustering by examining all possible partitions of n documents of t-dimensions into K clusters is not computationally feasible. Obviously, we need to resort to some optimization techniques to reduce the search space, but then there is no guarantee that the optimal solution will be found. Methods such as GA [36, 41], self-organizing maps (SOM) [42] and ant clustering [43] have been used for document clustering. Particle swarm optimization (PSO) [44] is another computational intelligence method that has been applied to image clustering and other low dimensional datasets in [39, 45, 46] and to document clustering in [42]. HS is employed for document clustering in [47, 48].

To compare the quality and the speed of different clustering algorithms, some known data sets are available and have been used. In all of datasets, before applying clustering algorithm, the very common words (stop words) are stripped out completely and different forms of a word are reduced to one canonical form by using Porter's algorithm and then converted to the vector space model.

To demonstrate the document clustering accuracy in comparison to the best contemporary methods, five data sets are selected from different known sources. Data sets DS1 and DS2 are from TREC-5, TREC-6, and TREC-7 [49]; the data set DS3 was derived from the San Jose Mercury newspaper articles that are distributed as part of the TREC collection (TIPSTER); the data set DS4 is selected from the DMOZ collection; and the DS5 dataset is a collection of 10,000 messages, collected from 10 different Usenet newsgroups (1,000 messages from each). After preprocessing, there are a total of 9249 documents in this data set.

Figure 8 compares five different algorithms on the selected datasets. These algorithms includes HS clustering [47], *K*-means (best known partitioning algorithm), genetic *K*-means (GA) [50], particle swarm optimization based clustering (PSO) [42] and a Mises-Fisher generative model based algorithm (GM) [51, 52]. Figure 8 shows the results of applying these algorithms on five datasets considering the normalized ADDC of algorithm. From the results, it is easy to know that the HS method outperforms GA, *K*-means, and PSO in all datasets, while the GM algorithm generates higher quality clusters than the HS based algorithm for the dataset DS2.

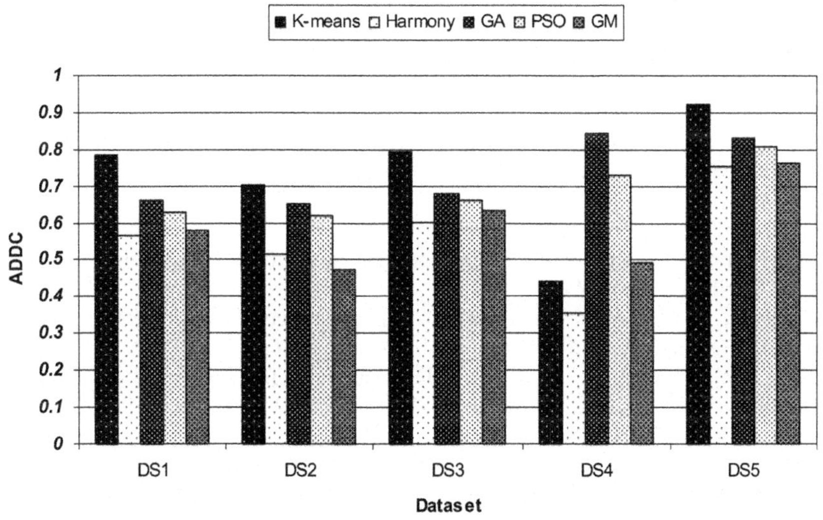

Fig. 8. Quality of clustering generated by various algorithms

8 Conclusions

This chapter reviewed HS applications in solving various NP-complete problems in computer science including the Sudoku puzzle, music composition, generalized orienteering problem, vehicle routing, multicast routing, and document clustering. The computational results showed that HS has had comparative advantages to other metaheuristic algorithms in all of these problems and the HS algorithm possesses a potential for obtaining good solutions in a variety of optimization problems.

For future study, one of the most important research directions is the theoretical analysis of HS. Since there was no research on the convergence of the HS algorithm, such work will have a significant effect on the application of HS. Also, novel methods to modify a pitch adjustment process when applying it to discrete-valued problems (i.e., combinatorial optimization problems) are the subject of more research. Last but not least, with the above-mentioned successful applications, HS has shown its capability to be applied to other NP-complete problems.

References

1. Geem, Z.W.: Harmony search algorithm for solving sudoku. In: Apolloni, B., Howlett, R.J., Jain, L. (eds.) KES 2007, Part I. LNCS (LNAI), vol. 4692, pp. 371–378. Springer, Heidelberg (2007)
2. Nicolau, M., Ryan, C.: Solving sudoku with the gAuGE system. In: Collet, P., Tomassini, M., Ebner, M., Gustafson, S., Ekárt, A. (eds.) EuroGP 2006. LNCS, vol. 3905, pp. 213–224. Springer, Heidelberg (2006)
3. Horner, A., Goldberg, D.E.: Genetic algorithms and computer-assisted music composition. In: Proceedings of the International Computer Music Conference, pp. 437–441 (1991)
4. Ralley, D.: Genetic algorithms as a tool for melodic development. In: Proceedings of the International Computer Music Conference, pp. 501–502 (1995)
5. Biles, J.A.: GenJam in perspective: a tentative taxonomy for GA music and art systems. Leonardo 36, 43–45 (2003)
6. Geem, Z.W., Choi, J.-Y.: Music composition using harmony search algorithm. In: Giacobini, M. (ed.) EvoWorkshops 2007. LNCS, vol. 4448, pp. 593–600. Springer, Heidelberg (2007)
7. Wang, Q., Sun, C., Golden, B.L.: Using artificial neural networks to solve generalized orienteering problems. In: Proceedings of Artificial Neural Networks in Engineering Conference (ANNIE 1996) (1996)
8. Chao, I.M., Golden, B.L., Wasil, E.A.: A fast and effective heuristic for the orienteering problem. European Journal of Operational Research 88, 475–489 (1996)
9. Geem, Z.W., Tseng, C.-L., Park, Y.-J.: Harmony search for generalized orienteering problem: Best touring in china. In: Wang, L., Chen, K., S. Ong, Y. (eds.) ICNC 2005. LNCS, vol. 3612, pp. 741–750. Springer, Heidelberg (2005)
10. Geem, Z.W., Lee, K.S., Park, Y.: Application of harmony search to vehicle routing. American Journal of Applied Sciences 2, 1552–1557 (2005)
11. Koumousis, V.K., Georgiou, P.G.: Genetic algorithms in discrete optimization of steel truss roofs. ASCE J. Comp. in Civil Eng. 8, 309–325 (1994)
12. Kompella, V.P., Pasquale, J.C., Polyzos, G.C.: Multicast routing for multimedia communication. IEEE/ACM Transactions on Networking 1, 286–292 (1993)

13. Zhu, Q., Parsa, M., Garcia-Luna-Aceves, J.J.: A source based algorithm for delay-constrained minimum-cost multicasting. In: Proceedings of the fourteenth annual joint conference of the IEEE computer and communication societies, pp. 377–385 (1995)
14. Widyono, R.: The design and evaluation of routing algorithms for realtime channels. Technical report TR-94-024, University of California at Berkeley (1994)
15. Raghavan, S., Manimaran, G., Ram Murthy, C.S.: A re-arrangeable algorithm for the construction of delay-constrained dynamic multicast trees. IEEE/ACM Transactions on Networking 7, 514–529 (1999)
16. Salama, H.F., Reeves, D.S., Viniotis, Y.: Evaluation of multicast routing algorithms for real-time communication on high-speed networks. IEEE Journal on Selected Areas in Communications 15, 332–345 (1997)
17. Kun, Z., Yong, Q., Hong, Z.: Dynamic multicast routing algorithm for delay and delay variation-bounded Steiner tree problem. Knowledge-Based Systems 19, 554–564 (2006)
18. Haghighat, A.T., Faez, K., Dehghan, M., Mowlaei, A., Ghahremani, Y.: GA-based heuristic algorithms for bandwidth-delay-constrained least-cost multicast routing. Computer Communications 27, 111–127 (2004)
19. Galiasso, P., Wainwright, R.L.: A hybrid genetic algorithm for the point to multipoint routing problem with single split paths. In: Proceedings of ACM Symposium on Applied Computting, pp. 327–332 (2001)
20. Ericsson, M., Resende, M., Pardalos, J.M.: A genetic algorithm for the weight setting problem in OSPF routing. Journal of Combinatorial Optimization 6, 299–333 (2002)
21. Barolli, L., Koyama, A., Suganuma, T., Shiratori, N.: A GA-based QoS routing method for mobile Ad-Hoc networks. Journal of Interconnection Networks 4, 251–270 (2003)
22. Zhang, Q., Lenug, Y.W.: An orthogonal genetic algorithm for multimedia multicast routing. IEEE Transactions on Evolutionary Computation 3, 53–62 (1999)
23. Wang, X., Cao, J., Cheng, H., Huang, M.: QoS multicast routing for multimedia group communications using intelligent computational methods. Computer Communications 29, 2217–2229 (2006)
24. Skorin-Kapov, N., Kos, M.: The application of Steiner trees to delay constrained multicast routing: a Tabu Search approach. In: Proceedings of the seventh international conference on telecommunications, pp. 443–448 (2003)
25. Prim, R.: Shortest connection networks and some generalizations. Bell System Technical Journal 36, 1389–1401 (1957)
26. Wang, H., Wang, J., Wang, H., Sun, Y.: TSDLMRA: an efficient multicast routing algorithm based on tabu search. Journal of Network and Computer Applications 27, 77–90 (2004)
27. Zhang, B., Mouftah, H.: A destination-driven shortest path tree algorithm. In: Proceedings of the IEEE international conference on communications, pp. 2258–2262 (2002)
28. Yen, J.Y.: Finding the K-shortest loop-less paths in a network. Management Science 17, 712–716 (1971)
29. Glover, F., Laguna, M.: Tabu search. Kluwer Academic, Dordrecht (1997)
30. Wang, Z., Shi, B., Zhao, E.: Bandwidth-delay-constrained least-cost multicast routing based on heuristic genetic algorithm. Computer Communications 24, 685–692 (2001)
31. Forsati, R., Mahdavi, M., Haghighat, M.T.: Bandwidth delay constrained least cost multicast routing. Computer Communications 31, 2505–2519 (2008)
32. Forsati, R., Mahdavi, M., Haghighat, M.T.: An efficient algorithm for bandwidth delay constraint least cost multicast routing. In: Proceedings of IEEE Canadian Conference on Electrical and Computer Engineering, pp. 1641–1646 (2008)

33. Wang, Z., Crowcroft, J.: Quality of service for supporting multimedia applications. IEEE Journal on Selected Areas in Communications 14, 1228–1234 (1996)
34. Jain, A.K., Murty, M.N., Flynn, P.J.: Data clustering: a review. ACM Computing Surveys (CSUR), 264–323 (1999)
35. Grira, N., Crucianu, M., Boujemaa, N.: Unsupervised and semi-supervised clustering: a brief survey. In: Proceedings of 7th ACM SIGMM international workshop on Multimedia information retrieval, pp. 9–16 (2005)
36. Raghavan, V.V., Birchand, K.: A clustering strategy based on a formalism of the reproductive process in a natural system. In: Proceedings of the Second International Conference on Information Storage and Retrieval, pp. 10–22 (1979)
37. Everitt, B.: Cluster analysis, 2nd edn. Halsted Press, New York (1980)
38. Salton, G., Buckley, C.: Term-weighting approaches in automatic text retrieval. Information Processing and Management 24, 513–523 (1988)
39. Cios, K., Pedrycs, W., Swiniarski, R.: Data mining methods for knowledge discovery. Kluwer Academic Publishers, Dordrecht (1998)
40. Salton, G.: Automatic text processing: the transformation, analysis, and retrieval of information by computer. Addison-Wesley, Reading (1989)
41. Jones, G., Robertson, A.M., Santimetvirul, C., Willett, P.: Non-hierarchic document clustering using a genetic algorithm. Information Research 1 (1995)
42. Cui, X., Potok, T.E., Palathingal, P.: Document clustering using particle swarm optimization. In: Proceedings of IEEE Swarm Intelligence Symposium, pp. 185–191 (2005)
43. Labroche, N., Monmarche, N., Venturini, G.: AntClust: ant clustering and web usage mining. In: Cantú-Paz, E., Foster, J.A., Deb, K., Davis, L., Roy, R., O'Reilly, U.-M., Beyer, H.-G., Kendall, G., Wilson, S.W., Harman, M., Wegener, J., Dasgupta, D., Potter, M.A., Schultz, A., Dowsland, K.A., Jonoska, N., Miller, J., Standish, R.K. (eds.) GECCO 2003. LNCS, vol. 2723, pp. 25–36. Springer, Heidelberg (2003)
44. Kennedy, J., Eberhart, R.C., Shi, Y.: Swarm intelligence. Morgan Kaufmann, New York (2001)
45. Omran, M., Salman, A., Engelbrecht, A.P.: Image classification using particle swarm optimization. In: Proceedings of the 4th Asia-Pacific Conference on Simulated Evolution and Learning, pp. 370–374 (2002)
46. Merwe, V.D., Engelbrecht, A.P.: Data clustering using particle swarm optimization. In: Proceedings of IEEE Congress on Evolutionary Computation, pp. 215–220 (2003)
47. Mahdavi, M., Chehregani, M.H., Abolhassani, H., Forsati, M.: Novel meta-heuristic algorithms for clustering web documents. Appl. Math. Comput. 201, 441–451 (2008)
48. Forsati, R., Mahdavi, M., Kangavari, M.: Web page clustering using harmony search optimization. In: Proceedings of IEEE Canadian Conference on Electrical and Computer Engineering, pp. 1601–1604 (2008)
49. Text REtrieval Conference (1999) (accessed November 30, 2008), `http://trec.nist.gov`
50. Krishna, K., Murty, M.N.: Genetic K-means algorithm. IEEE Transactions on Systems, Man and Cybernetics (Part B. Cybernetics) 29, 433–439 (1999)
51. Zhong, S., Ghosh, J.: Generative model-based clustering of documents: a comparative study. Knowledge and Information Systems 8, 374–384 (2005)
52. Some Software Packages (2005) (accessed November 30, 2008), `http://www.cse.fau.edu/~zhong/software/index.htm`

Harmony Search Applications in Mechanical, Chemical and Electrical Engineering

Mohammad Fesanghary

Department of Mechanical Engineering, Amirkabir University of Technology, Tehran, Iran
fesanghary@gmail.com

Abstract. The primary aim of this chapter is to introduce the state of the art applications of the harmony search (HS) algorithm in mechanical, chemical and electrical engineering fields. The HS algorithm has been broadly utilized in complex optimization problems arising in most engineering applications. It has been reported to be a viable alternative to other conventional optimization techniques in these engineering disciplines. This chapter is intended to discuss the available literature in the field and to provide general information for those who are interested in optimized-design of thermal systems, economic utilization of electric power-systems and optimization of machining processes.

Keywords: Harmony Search, Heat Exchanger Design, Economic Dispatch, Machining Process.

1 Introduction

This chapter deals with harmony search applications in mechanical, chemical and electrical engineering. Various engineering optimization problems such as design optimization of heat exchangers, economic utilization of electric power systems and selection of cutting parameters in machining processes will be discussed in here.

A heat exchanger is a device built for efficient heat transfer from one medium to another. The media can either be separated by a solid wall so that they never mix, or be in direct contact. Heat exchangers are widely used in space heating, refrigeration, air conditioning, power plants, chemical plants, petrochemical plants, petroleum refineries, and natural gas processing. This chapter shows the HS application on some heat exchanger design problems including design of shell and tube heat exchangers (STHEs), air-cooled heat exchangers (ACHEs) and satellite heat pipes.

The HS algorithm is also applied to power economic dispatch (ED) problems. Economic dispatch is considered to be one of the key functions in electric power system operation. The ED problem is commonly formulated as an optimization problem aimed to minimize the total generation cost of units, while satisfying generation/network constraints. In order to show the performance of the HS algorithm in solving power dispatch problems a benchmark problem will be considered.

Another HS application in mechanical engineering field is optimization of machining processes. Machining is one of the most important material removal methods used in industry and is a part of the manufacture of almost all metal products. The production cost and quality of machined parts are determined by many variables such as cutting speed, feed, and depth of cut. This chapter shows how efficiently these parameters can be optimized using HS algorithm.

Z.W. Geem (Ed.): Music-Inspired Harmony Search Algorithm, SCI 191, pp. 71–86.
springerlink.com

2 Design Optimization of Heat Exchangers

Design optimization of Heat exchangers is a complex task due to the coupled nature of the design parameters involved. This section will focus on design optimization of heat exchangers from an economic point of view and review some real-world examples.

2.1 Shell and Tube Heat Exchangers

Shell and tube heat exchangers (STHE) are the most common types of heat exchangers in oil refineries and other large chemical processes, and are suited for a wide range of operating temperatures and pressures. This type of heat exchanger consists of a shell with a bundle of tubes inside. One fluid runs through the tubes and another fluid flows over the tubes (through the shell) to transfer heat between the two fluids [1]. A typical STHE is illustrated in Fig. 1. From a design point of view, STHEs represent a complex process containing an integrated whole of design rules and empirical knowledge of various fields. In fact, there are several parameters, such as tube diameter, tube length, tube pitch, tube pitch pattern, tube passes, baffle cuts, baffle spacing, and tube material, which need to be taken into account in design of STHEs.

Similar to most thermal systems, different types of variables including real, integer and discrete types take part in design problem of heat exchangers. Integer variables appear when the quantity of some identical components is a variable, such as the number of tubes or tube passes in heat exchangers. Discrete variables usually arise from discrete standard sizes of materials, such as the standard tube diameter, shell diameter and tube pitch pattern.

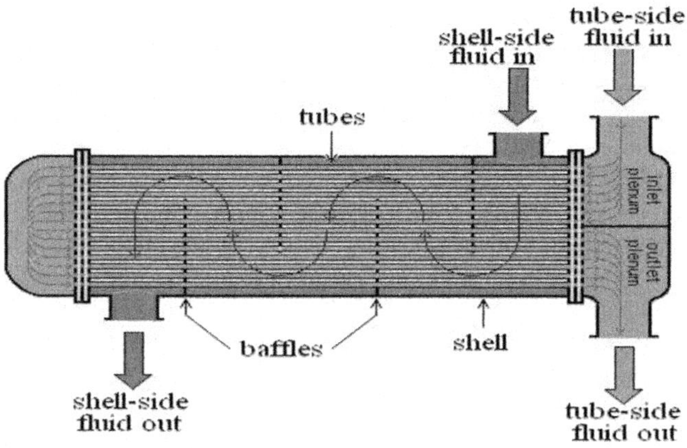

Fig. 1. Schematic of a shell and tube heat exchanger

2.1.1 Problem Formulation for Design of STHEs

The objective function in heat exchanger design problems is normally the total annual cost of the exchanger which includes the capital cost and pumping costs. However, there are some studies in the literature which consider the rate of entropy generation as the objective function as well [2-4]. Capital cost is usually approximated based on the surface area of tubes using experimental formulas and pumping cost are calculated based on electricity fees and pressure drop of fluids in the exchanger. Calculation of the required surface area and pressure drop of fluids is a very complex process which can be found in much detail in heat transfer textbooks [5-7].

There is a major design constraint which should be satisfied in a practical design. The STHE should be capable of accomplishing the prescribed heat load. Also, there are many other thermal/mechanical/practical limitations which should be considered. For example when designing the tubes, it is practical to ensure that the tube pitch is not less than 1.25 times the tubes' outside diameter, the baffle cut which is the ratio of the baffle window height to the shell diameter should be greater than 15%, and baffles should not be spaced closer than about one-fifth of the shell diameter. Furthermore, because of fouling problems and mechanical cleaning limitations, tube diameters should be greater than a specified value.

2.1.2 Example of STHE Design

This example taken from Fesanghary et al. [8] is concerned with the design of a shell and tube oil-cooler. The light oil enters the exchanger at 102°C and is to be cooled down to 64°C by water entering at 21°C. The objective is to find the most economic design which can satisfy the heat load of the exchanger together with pressure drop limitations.

To show the ability of the HS method, its performance is compared with a genetic algorithm (GA). Although both GA and HS have proven their abilities in finding near global solutions within a reasonable amount of time, they are comparatively inefficient in finding the precise optimum solution. To evaluate the accuracy of the obtained results, the global optimum solution is also calculated considering all possible exchanger geometries. Thus, first the continuous variables are divided to 100 equal sections in their range of variations. Considering discrete variables, a total number of 67,797,000,000 combinations are obtained. Then all the candidate solutions are examined to find the optimal solution. Table 1 lists the comparative performances for the exchanger cost and average central processing unit (CPU) times for HS and GA.

Table 1. Cost comparison for STHE design example

Algorithms	Total Cost	CPU time (sec.)
GA	$10630.5	13.12
HS	$10572.0	8.38
Global optimum [8]	$10553.4	591,453

Results reveal that both HS and GA can converge to near optimal solutions in a reasonable amount of time. The difference between the global optimum solution and those obtained using HS is 0.2 percent. This difference becomes 0.7 percent for GA. Note that to find the global optimal solution, the time elapsed through evaluating all possible combinations was about 164 hours while it took only a few seconds for HS to find a good near optimal solution.

2.2 Air-Cooled Heat Exchangers

Atmospheric air has been used for many years to cool and condense fluids in areas of water scarcity. During the 1960s, the use of ACHEs grew rapidly worldwide. In Europe, where seasonal variation in ambient temperature is relatively small, ACHEs are used for the greater part of cooling process [9]. ACHEs are increasingly being used nowadays due to significant increase in water cost, lack of water supplies, and concern for water pollution.

A typical ACHE includes a tube bundle, which generally has spiral-wound fins upon the tubes, and a fan, which moves air across the tubes and is provided with a driver. Electric motors are the most commonly used drivers. A typical ACHE is shown in Fig.2.

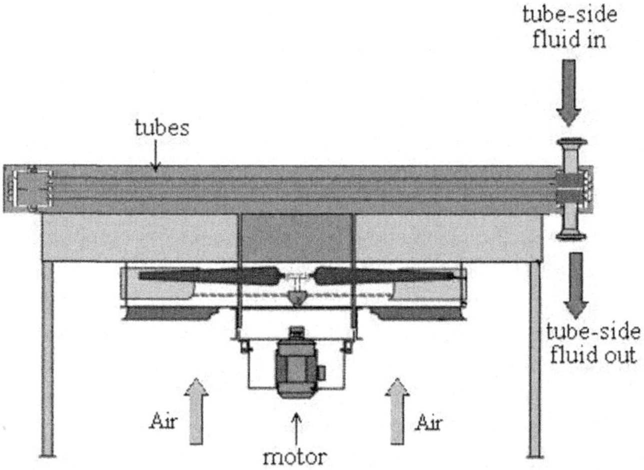

Fig. 2. Schematic of an air-cooled heat exchanger

Compared to the shell and tube exchangers, however, there are more parameters to be taken into account, with the most important ones being tube diameter, tube length, tube rows, number of tubes per row, air velocity, fan type, fan diameter, number of fins, fin height, tube pattern and tube material. Similar to the shell and tube heat exchangers, an ACHE design problem has different variable types. Tube rows, number of tubes per row and number of fins are integer variables, whereas tube diameter, fin height, and tube pattern are discrete variables and can be chosen only from prespecified standard values.

In the last few decades, several methods for optimization of heat exchangers have been developed. These methods can be classified into three main groups, namely thermodynamic approaches, mathematical programming methods and stochastic methods. Thermodynamic approaches are based on Second Law Analysis. The aim of these approaches is to minimize the exergy destruction or irreversibility. Mathematical programming methods have been used for design optimization of heat exchangers in the last decade [10-14]. Although accurate and fast, these approaches rely strictly on the initial starting point, the topology of the feasible region and the surface associated with the objective function. In other words, a good starting point is crucial for these methods to function successfully. These drawbacks therefore decrease the effectiveness of these methods in thermal system design applications where the problems are usually non-convex and have a large amount of discrete design variables. Stochastic methods perform well for global searching and have the capability of quickly exploring and finding high performance regions in the search space. These methods are able to handle complex problems without being limited by the nonlinearities, the non-convexities and the discontinuities of the objective function [15]. Due to these capabilities, there has been a trend recently among researchers to implement these methods to heat exchanger design optimization problems. Simulated annealing [16] and genetic algorithms (GAs) [17-22] are the most frequently used methods among this certain class of optimization methods.

2.2.1 Problem Formulation for Design of ACHEs

The objective is to find the optimal design of an ACHE which is capable of accomplishing the prescribed thermal duty with minimum total cost. The overall cost associated with a heat exchanger may be categorized as the capital and operating costs. The capital cost includes the cost associated with design, materials, manufacturing, testing, shipping and installation, which is usually estimated based on empirical formulas expressed in terms of tube surface area. On the other hand, the operating cost consists of the expense associated with electrical power consumed by fans.

There are many thermal/mechanical/practical limitations which should be considered in a practical design. Each design parameter has a specified/recommended range of variation. Besides this, other constraints such as energy balance, mass conservation, and maximum allowable pressure drop and fluid velocity in tubes should be satisfied in the optimization process.

2.2.2 Example of ACHE Design

As an illustrative example, a refrigeration air-cooled condenser is considered here. Process condition used for this problem is given in [23]. To show the performance of the HS its results are validated by means of comparison with GA as shown in Table 2.

Table 2. Cost comparison for ACHE design example

Algorithms	Total Cost	CPU time (sec.)
GA	\$2627.0	11.43
HS	\$2620.9	2.14

It can be seen that HS shows better performance than GA in this case. Although the cost of the ACHEs obtained by GA and HS are close to each other, the geometries obtained by these algorithms are completely different. Also, the computational time elapsed in HS method is comparatively less than that of GA approach.

2.3 Satellite Heat Pipes

A heat pipe is a heat transfer device that can transport large quantities of heat between two interfaces that have very small temperature difference. Heat pipes are used widely in air-conditioners, refrigerators, capacitors, transistors and etc. They have also a significant application in the field of cryogenics, especially in the development of space technology [24]. Until recently, the use of heat pipes has mainly been limited to the space technology due to cost effectiveness and complex wick construction of heat pipes. The main applications of heat pipes in the space technology are [25]:

- Temperature equalization of spacecraft.
- Component cooling and temperature control in satellites.
- Removal of heat from the reactor and elimination of troublesome thermal gradients along the collector in spacecrafts.

A schematic of satellite heat pipe is illustrated in Fig. 3. Five important parameters including length of flange fin (L_f), cutting length of adhesive attached area (L_c), thickness of fin (t_f), adhesive thickness (t_b), and operation temperature (T_{op}) are considered as design variables.

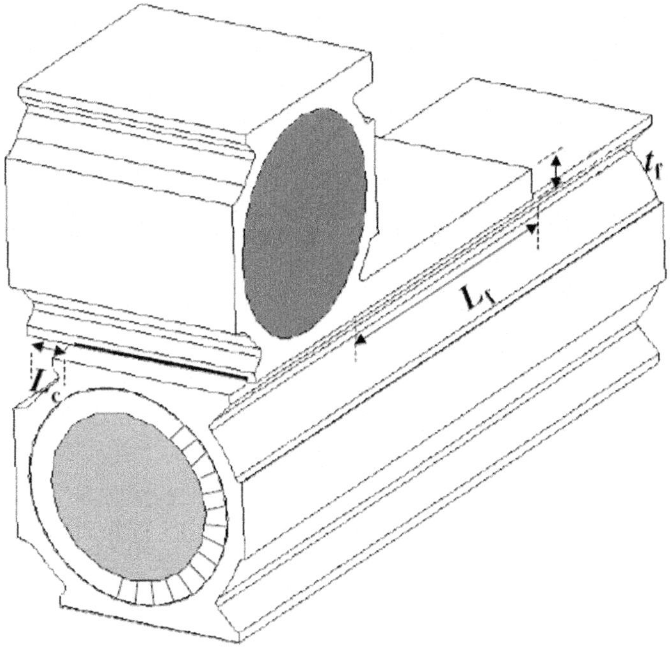

Fig. 3. Schematic of a satellite heat pipe

2.3.1 Example of Satellite Heat Pipe Design

Geem and Hwangbo [26] considered the design of a satellite heat pipe whose objectives were minimization of the heat pipe mass and maximization of thermal conductance.

This problem was formulated as an unconstrained, multiple-objective optimization problem having five continuous design variables:

$$\min \quad Z = \left| \frac{G(L_f, L_c, t_f, t_b, T_{op}) - G^*}{G^*} \right| + \left| \frac{M(L_f, L_c, t_f, t_b) - M^*}{M^*} \right| \tag{1}$$

where G is the thermal conductance; M is the mass; and G^* and M^* are the optimal values of G and M determined by optimizing each function separately. The result obtained using HS is compared with those of Broyden-Fletcher-Goldfarb-Shanno (BFGS) [27] in Table 3.

Table 3. Optimal results for satellite heat pipe design example

Algorithms	G (W/K)	M (kg)
BFGS	0.3750	26.854
HS	0.3810	26.704

As seen from Table 3, HS can find better solutions compared to those of gradient-based technique BFGS. The mass obtained by HS is 0.6% lower while thermal conductance by HS is 1.6% higher, when compared with BFGS.

3 Economic Power Dispatch

Economic dispatch is the method of determining the most efficient, low-cost and reliable operation of a power system by dispatching the available electricity generation resources to supply the load on the system. The primary objective of economic dispatch is to minimize the total cost of generation while honoring the operational constraints of the available generation resources [28].

Various mathematical programming methods such as linear programming, homogenous linear programming, and nonlinear programming [29-34] have been applied so far for solving the ED problems. These methods use gradient information to search solution space near an initial starting point. In general, gradient-based methods converge faster and can obtain solutions with higher accuracy compared to stochastic approaches in fulfilling the local search task. However, for effective implementation of these methods, the variables and cost function of the generators need to be continuous. Furthermore, a good starting point is crucial for these methods to function successfully. In a realistic operation, to provide completeness for the ED problem formulation, prohibited operating zones, ramp-rate limits, and non-smooth or non-convex cost functions arising due to the use of multiple fuels should be considered. The resulting ED formulation is a non-convex optimization problem, which is a challenging one and cannot be solved by the traditional mathematical programming methods. Therefore, dynamic programming [35] and mixed integer nonlinear programming [36], and their modifications have been presented in the literature. However, these methods may cause the dimensions of the ED problem to become extremely large when applied on large power systems, thus requiring enormous computational efforts [37].

Recently, as an alternative to the conventional mathematical approaches, the meta-heuristic optimization techniques such as genetic algorithm [37-39], Tabu search [40], simulated annealing (SA) [41], ant colony (AC) [42], evolutionary programming (EP) [43] and particle swarm optimization (PSO) [44-46] have been used to obtain global or near global optimum solutions for ED problems. These methods are effective for global searching due to their capability of exploring and finding high performance regions in the search space at an affordable time. Also, limitations regarding the form of the cost functions employed and the continuity of variables used for the mathematical optimization methods can be completely eliminated.

Although various optimization methodologies have been developed for economic dispatch problems, the complexity of the task reveals the need for developing efficient algorithms to precisely locate the optimum dispatch solution. In this context, this chapter focuses on the application of the HS algorithm for solving ED problems, aiming to provide a practical alternative for conventional methods.

3.1 Problem Formulation for ED Problems

The goal of the ED problem is to minimize the total power generation cost that is modeled as the sum of the cost functions of all generators. To simplify the optimization problem and facilitate the application of classical techniques, cost functions of generators are typically modeled by smooth quadratic function form given as:

$$C_T = \sum_{i=1}^{n} a_i + b_i P_i + c_i P_i^2 \tag{2}$$

where C_T is the total generation cost, n is the total number of generating units, a_i, b_i and c_i are the cost coefficients of the ith unit, and P_i is the actual power output of the ith unit.

To model a more realistic cost function of generators, the valve-point effects need to be considered. For the purpose of modeling the valve-point loading effects a recurring rectified sinusoidal term is added to Eq. (2) as follows:

$$C_T = \sum_{i=1}^{n} a_i + b_i P_i + c_i P_i^2 + \alpha \left| d_i \sin(f_i (P_i^{\min} - P_i)) \right| \tag{3}$$

where d_i and f_i are the valve-point coefficients of the ith unit. $P_i^{\min}$ is the lower generation limit of the ith unit.

Practically several types of fuel may be used for a unit. For a unit with k fuel options, Eq. (2) is modified as follows.

$$C_{Ti} = \begin{cases} a_{i1} + b_{i1}P_i + c_{i1}P_i^2 & fuel\,1 \quad P_i^{\min} \le P_i \le P_{i1} \\ a_{i2} + b_{i2}P_i + c_{i2}P_i^2 & fuel\,2 \quad P_{i1} < P_i \le P_{i2} \\ \vdots & \\ a_{ik} + b_{ik}P_i + c_{ik}P_i^2 & fuel\,k \quad P_{ik-1} < P_i \le P_{ik} \end{cases} \tag{4}$$

where C_{Ti} is the fuel cost function of the ith unit. The coefficients a_{ik}, b_{ik} and c_{ik} are cost coefficients of the ith unit using fuel type k.

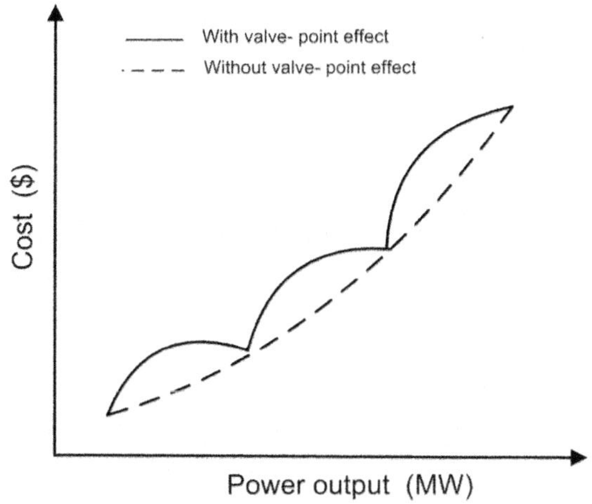

Fig. 4. Valve-point effects

Generally, the ED problem is subjected to the following constraints:

(i) Power balance constraint

$$\sum_{i=1}^{n} P_i = P_D + P_L \tag{5}$$

where P_D is the total demand and P_L is the total transmission network loss.

(ii) Generation limit constraint
Generation power of each generator is assumed between maximum and minimum limits:

$$P_i^{\min} \le P_i \le P_i^{\max} \tag{6}$$

(iii) Prohibited zones constraint
A unit with prohibited operating zones has a discontinuous input–output power generation characteristic (Fig. 5), and as a result, there are additional constraints on the unit operating range:

$$P_i \in \begin{cases} P_i^{\min} \le P_i \le P_{i,1}^{l} \\ P_{i,k-1}^{u} \le P_i \le P_{i,k}^{l} \quad , k = 2,\ldots,z_i \\ P_{i,z_i}^{u} \le P_i \le P_i^{\max} \end{cases} \tag{7}$$

where $P_{i,k}^{l/u}$ is the lower/upper bound of the kth prohibited zone of unit i and z_i is the ith prohibited zone.

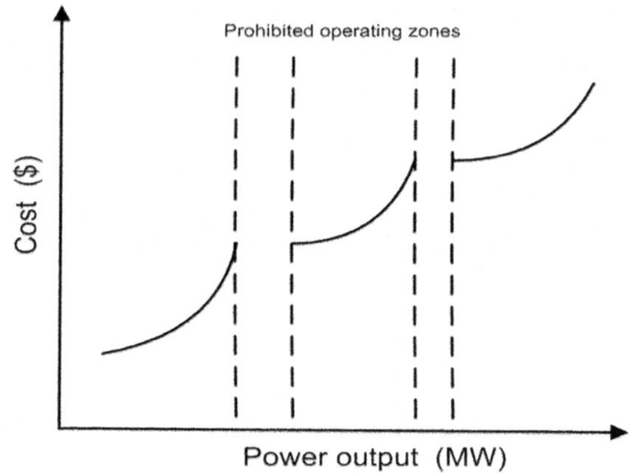

Fig. 5. Prohibited operating zones

(iv) Line flow constraint
Line flow constraints can be expressed as follows:

$$|Lf_i| \le Lf_i^{\max}, \quad i = 1..N_L \tag{8}$$

where Lf_i is the MW line flow. $Lf_i^{\max}$ is the allowable maximum flow of transmission line i (line capacity), and N_L is the number of transmission lines subject to line capacity constraints.

(v) Ramp-rate limits constraint
To avoid excessive thermal stresses on the boiler and combustion equipments, the rate of change in the output power of each thermal unit must not exceed a certain rate during increasing or decreasing the power output of each unit. This constraint can be formulated as follows:

$$\max(P_i^{\min}, P_i^{o} - DR_i) \le P_i \le \min(P_i^{\max}, P_i^{o} + UR_i) \tag{9}$$

where P_i^{o} is the previous output power. UR_i is the upramp limit of the ith generatorand and DR_i is the downramp limit of the ith generator.

(vi) Feasible operating region
In combined heat and power (CHP) units the power capacity limits of units are functions of the unit heat productions. Respectively the heat capacity limits are functions of the unit power generations. Usually these constraints are expressed by a closed boundary curve as shown in Fig. 6 (i.e. curve ABCDEFA).

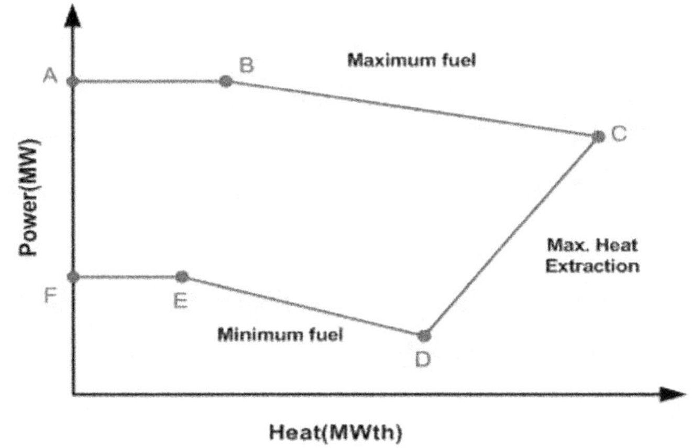

Fig. 6. Feasible operation region for a typical cogeneration unit

3.2 Example of ED Problem

One of the most challenging ED problems is emerged in CHP systems. Cogeneration systems have now been extensively utilized by the industry, because they offer a reliable, efficient and economic mean to supply both thermal and electrical energy. To show the effectiveness of the HS algorithm in solving ED problems, a benchmark problem is considered here.

This problem was originally proposed by Gou et al. [47]. The problem includes a conventional power unit, two cogeneration units and a heat-only unit. The objective is to find the minimum overall cost of units subject to constraints on heat and power production. Table 4 presents the best solution of this problem obtained using the HS algorithm and compares the HS results with those obtained from improved genetic algorithm (IGA) [51], genetic algorithm [48], Lagrangian Relaxation (LR) [47], ant colony search algorithm (ACSA) [49] and a hybrid of genetic algorithm with tabu search (GT) [50].

Table 4. Optimal results for ED problem example

Algorithms	Minimum Cost
LR [47]	$9257.10
GA [48]	$9267.20
ACSA [49]	$9452.20
GT [50]	$9207.64
IGA_ MU [51]	$9257.07
HS	$9186.97

It is obvious from the Table 1 that the result obtained by HS algorithm is more economic than those reported previously in the literature.

4 Machining Process

There are many kinds of machining operations such as turning, drilling, boring and milling, each of them being capable of generating a certain part geometry and surface texture. To provide comprehensive information for optimization of machining operations in a reasonable length, the detailed description of all methods is not presented here, as these can be found in machining textbooks [52]. Instead, we focus only on the milling process in this section.

In milling, a rotating tool with multiple cutting edges is moved slowly relative to the material to generate a plane or straight surface (Fig. 7). The primary motion is accomplished at a certain cutting speed. In addition, the tool must be moved laterally across the work. This is a much slower motion, called the feed. The remaining dimension of the cut is the penetration of the cutting tool below the original work surface, called the depth of cut. Collectively, speed, feed, and depth of cut are called the cutting parameters.

The cost, time and quality of production are highly dependent on the cutting parameters. Thus, determination of optimal cutting parameters with regard to technological requirements, capability of machine tool, cutting tool and the part material is a crucial task in the process planning of parts. Several approaches such as feasible direction method [53], dynamic programming [54], geometric programming [55], and genetic algorithm [56, 57] have been introduced so far for the optimization of cutting parameters. This section will focus on the application of HS for the optimization of cutting parameters.

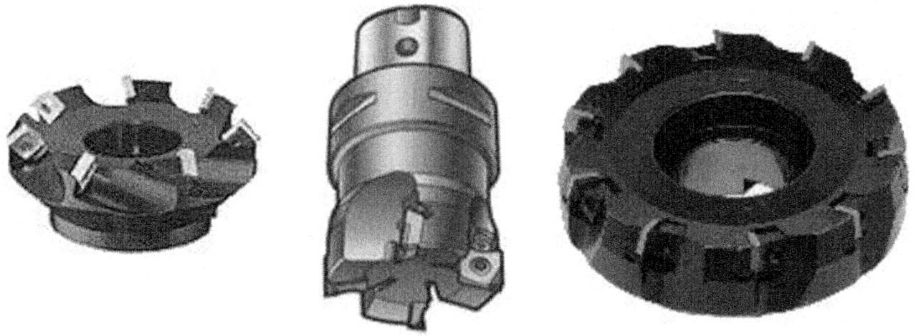

Fig. 7. Typical milling tools

4.1 Problem Formulation for Milling Process

The machining operation can be divided into the roughing and finishing operations. In the roughing operation a large amount of material from the starting workpart is removed as rapidly as possible in order to produce a shape close to the desired form, but some material is left on the piece for a subsequent finishing operation. Finishing cuts are used to complete the part and achieve the final dimension, tolerances, and surface smoothness. In production machining jobs, one or more roughing cuts are usually performed on the work, followed by one or two finishing cuts. The total production cost (C_T) for multi-pass milling is expressed as:

$$C_T = C_f + \sum_{i=1}^{n} C_{ri} + C_{tp} \tag{10}$$

where C_f is the cost of the finishing pass, C_{ri} is the cost of the ith roughing pass, and C_{tp} is the cost of tool preparation. The objective function, Eq. (10), is subjected to various constraints such as tool-life, available cutting speeds, surface roughness, and cutting power.

4.2 Example of Machining Process

An illustrative example is used to demonstrate the ability of the HS algorithm for solving machining optimization problems. For validation purpose, GA is also used to solve this problem. The data for this problem are available from Zarei et al. [58]. Table 5 shows the best HS and GA results for a total cutting depth of 8 mm.

Table 5. Optimal results for machining process example

Algorithms	Cost per pass			Total Cost
	1st	2nd	3rd	
GA	\$0.5366	\$0.3716	\$0.5047	\$1.4129
HS	\$0.4446	\$0.4446	\$0.5047	\$1.3939

Both HS and GA find the solutions with three passes (two rough passes and a finish pass). It can be seen from the results that the total cost obtained from HS is lower than GA.

5 Conclusions

Three types of engineering optimization problems were studied in this chapter. The first type is related to the design optimization of thermal systems; the second type concerns the economic utilization of electric power systems; and the third deals with the selection of cutting parameters in machining process. The ability of the HS algorithm was demonstrated using several test problems and its performance was compared with other conventional methods. The results reveal that HS outperforms other approaches not only in terms of the quality of the obtained solutions but also in terms of the elapsed computational time. Generally, it can be concluded that the HS simplicity of implementation, good computational efficiency, high quality solution, along with the lower number of setting parameters makes it an ideal method when dealing with complex engineering optimization problems. Future research can be directed towards HS implementation in diverse areas of mechanical/chemical/electrical engineering fields such as synthesis of heat exchanger networks, utilization of distributed generation sources (i.e. photovoltaic and pump storage generators) and more.

Acknowledgments

The author would like to thank Mr. R. Rabiei and M. Minary-Jolandan for their constructive comments and suggestions.

References

1. Wikipedia, Shell and tube heat exchanger (2005) (accessed October 25, 2008), `http://en.wikipedia.org/wiki/Shell_and_tube_heat_exchanger`
2. Özçelik, Y.: Exergetic optimization of shell and tube heat exchangers using a genetic based algorithm. Appl. Therm. Eng. 27, 1849–1856 (2007)
3. Sun, S.-y., Lu, Y.-d., Yan, C.-q.: Optimization in calculation of shell-tube heat exchanger. Int. Comm. Heat Mass Transfer 20, 675–685 (1993)
4. Lerou, P.P.P.M., Veenstra, T.T., Burger, et al.: Optimization of counterflow heat exchanger geometry through minimization of entropy generation. Cryogenics 45, 659–669 (2005)
5. Taborek, J.: Shell and tube heat exchangers: single phase flow, Heat Exchanger Design Handbook. Hemisphere, New York (1982)
6. Saunders, E.A.D.: Heat exchangers – selection, design, construction. Longman scientific and technical, New York (1988)
7. Hewitt, G.F., Shires, G.L., Bott, T.R.: Process heat transfer. CRC Press, Boca Raton (1994)
8. Fesanghary, M., Damangir, E., Soleimani, I.: Design optimization of shell and tube heat exchangers using global sensitivity analysis and harmony search algorithm. Appl Therm. Eng. (2008), doi:10.1016/j.applthermaleng.2008.05.018
9. Perry, R.J., Chilton, C.H.: Chemical Engineers Handbook. McGraw-Hill, New York (1973)
10. Reppich, M., Kohoutek, J.: Optimal design of shell-and-tube heat exchangers. Computers Chem. Engng. 18, 295–299 (1994)
11. Zalewski, W., Niezgoda-Zelasko, B., Litwin, M.: Optimization of evaporative fluid coolers. International Journal of Refrigeration 23, 553–565 (2000)
12. Gonzalez, M.T., Petrcci, N.C., Urbicain, M.J.: Air-cooled heat exchanger design using successive quadratic programming (SQP). Heat Transfer Engineering 22, 11–16 (2001)
13. Saffar-Avval, M., Damangir, E.: A general correlation for determining optimum baffle spacing for all types of shell and tube exchangers. Int. J. Heat Mass Transfer 38, 2501–2506 (1995)
14. Soltan, B.K., Saffar-Avval, M., Damangir, E.: Minimizing capital and operating costs of shell and tube condensers using optimum baffle spacing. Appl. Thermal Eng. 24, 2801–2810 (2004)
15. Fesanghary, M., Mahdavi, M., Minary, M., Alizadeh, Y.: Hybridizing harmony search algorithm with sequential quadratic programming for engineering optimization problems. Comput. Methods Appl. Mech. Engrg. 197, 3080–3091 (2008)
16. Chaudhuri, P.D., Diwekar, U.M., Logsdon, J.S.: An automated approach for the optimal design of heat exchangers. Ind. Eng. Chem. Res. 36, 3685–3693 (1997)
17. Selbaş, R., Kızılkan, Ö., Reppich, M.: A new design approach for shell-and-tube heat exchangers using genetic algorithms from economic point of view. Chemical Engineering and Processing 45, 268–275 (2006)

18. Ozkol, I., Komurgoz, G.: Determination of the optimum geometry of the heat exchanger body via a genetic algorithm. Numerical Heat Transfer, Part A 48, 283–296 (2005)
19. Gholap, A.K., Khan, J.A.: Design and multi-objective optimization of heat exchangers for refrigerators. Applied Energy 84, 1226–1239 (2007)
20. Allen, B., Gosselin, L.: Optimal geometry and flow arrangement for minimizing the cost of shell-and-tube condensers. Int. J. Energy Res. (2008), doi:10.1002/er.1398
21. Wildi-Tremblay, P., Gosselin, L.: Minimizing shell-and-tube heat exchanger cost with genetic algorithms and considering maintenance. Int. J. Energy Res. 31, 867–885 (2007)
22. Agarwal, A., Gupta, S.K.: Jumping gene adaptations of NSGA-II and their use in the multi-objective optimal design of shell and tube heat exchangers. Chemical Engineering Research and Design 86, 123–139 (2008)
23. Doodman, A., Fesanghary, M., Hosseini, R.: A robust stochastic approach for design optimization of air cooled heat exchangers. Applied Energy (2008), doi:10.1016/j.apenergy.2008.08.021
24. Wikipedia, heat pipe (2004) (accessed October 25, 2008), `http://en.wikipedia.org/wiki/Heat_pipe`
25. What is a Heat Pipe? (2008) (accessed October 25, 2008), `http://www.cheresources.com/htpipes.shtml`
26. Geem, Z.W., Hwangbo, H.: Application of harmony search to multi-objective optimization for satellite heat pipe design. In: Proceedings of the US-Korea Conference on Science, Technology, and Entrepreneurship (UKC 2006), CD-ROM (2006)
27. Rajesh, V.G., Ravindran, K.P.: Optimum heat pipe design: A nonlinear programming approach. International Communications in Heat and Mass Transfer 24, 371–380 (1997)
28. Wikipedia, economic dispatch (2005) (accessed October 25, 2008), `http://en.wikipedia.org/wiki/Economic_dispatch`
29. Lin, C.E., Chen, S.T., Huang, C.L.: A direct Newton-Raphson economic dispatch. IEEE Trans. Power Syst. 7, 1149–1154 (1992)
30. Lin, C.E., Viviani, G.L.: Hierarchical economic dispatch for piecewise quadratic cost functions. IEEE Trans. Power App. Syst. 103, 1170–1175 (1984)
31. Wood, A.J., Wollenberg, B.F.: Power generation, operation and control. Wiley, New York (1994)
32. Yang, H.T., Chen, S.L.: Incorporating a multi-criteria decision procedure into the combined dynamic programming/production simulation algorithm for generation expansion planning. IEEE Trans. Power Syst. 4, 165–175 (1989)
33. Granville, S.: Optimal reactive dispatch through interior point methods. IEEE Trans. Power Syst. 9, 136–146 (1994)
34. Chen, C.L., Wang, S.C.: Branch-and-bound scheduling for thermal generating units. IEEE Trans. Energy Convers. 8, 184–189 (1993)
35. Liang, Z.X., Glover, J.D.: A zoom feature for a dynamic programming solution to economic dispatch including transmission losses. IEEE Trans. Power Syst. 7, 544–550 (1992)
36. Papageorgiou, L.G., Fraga, E.S.: A mixed integer quadratic programming formulation for the economic dispatch of generators with operating zones. Electric Power Syst. Res. 77, 1292–1296 (2007)
37. Kuo, C.C.: A novel string structure for economic dispatch problems with practical constraints. Energy Convers. Manage (2008), doi:10.1016/j.enconman.2008.07.007
38. Li, F.: A comparison of genetic algorithms with conventional techniques on a spectrum of power economic dispatch problems. Expert Syst. Appl. 15, 133–142 (1998)
39. Baskar, S., Subbaraj, P., Rao, M.V.C.: Hybrid real coded genetic algorithm solution to economic dispatch problem. Comput. Electr. Eng. 29, 407–419 (2003)

40. Lin, W.M., Cheng, F.S., Tsay, M.T.: An improved tabu search for economic dispatch with multiple minima. IEEE Trans. Power Syst. 17, 108–112 (2002)
41. Basu, M.: A simulated annealing-based goal-attainment method for economic emission load dispatch of fixed head hydrothermal power systems. Int. J. Electron Power Energy Syst. 27, 147–153 (2005)
42. Song, Y.H., Chou, C.S.V., Min, Y.: Large-scale economic dispatch by artificial ant colony search algorithms. Electr. Mach. Power Syst. 27(7), 679–690 (1999)
43. Somasundaram, P., Kuppusamy, K., Kumudini Devi, R.P.: Economic dispatch with prohibited operating zones using fast computation evolutionary programming algorithm. Electr. Power Syst. Res. 70, 245–252 (2004)
44. Gaing, Z.L.: Particle swarm optimization to solving the economic dispatch considering the generator constraints. IEEE Trans. Power Syst. 18, 1187–1195 (2003)
45. Selvakumar, A.I., Thanushkodi, K.: Anti-predatory particle swarm optimization: Solution to nonconvex economic dispatch problems. Electr. Power Syst. Res. 78, 2–10 (2008)
46. Yuan, X., Su, A., Yuan, Y., et al.: An improved PSO for dynamic load dispatch of generators with valve-point effects. Energy (2008), doi:10.1016/j.energy.2008.09.010
47. Guo, T., Henwood, M.I., van Ooijen, M.: An algorithm for combined heat and power economic dispatch. IEEE Trans. Power Syst. 11, 1778–1784 (1996)
48. Song, Y.H., Xuan, Q.Y.: Combined heat and power economic dispatch using genetic algorithm based penalty function method. Electric Mach. Power Syst. 26, 363–372 (1998)
49. Song, Y.H., Chou, C.S., Stonham, T.J.: Combined heat and power economic dispatch by improved ant colony search algorithm. Electric Power Systems Research 52, 115–121 (1999)
50. Sudhakaran, M., Slochanal, S.M.R.: Integrating genetic algorithms and tabu search for combined heat and power economic dispatch. Power System Stability and Control 1, 67–71 (2003)
51. Su, C.T., Chiang, C.L.: An incorporated algorithm for combined heat and power economic dispatch. Electric Power Systems Research 69, 187–195 (2004)
52. Groover, M.P.: Fundamentals of Modern Manufacturing. Wiley, New York (2000)
53. Tolouei-Rad, M., Bidhendi, I.M.: On the optimization of machining parameters for milling operations. Int. J. Mach. Tools Manuf. 37, 1–16 (1997)
54. Sonmez, A.I., Baykasoglu, A., Dereli, T., et al.: Dynamic optimization of multipass milling operations via geometric programming. Int. J. Mach. Tools Manuf. 39, 297–320 (1999)
55. Jha, N.K.: A discrete data base multiple objective optimization of milling operation through geometric programming. Trans. ASME J. Eng. Ind. 112, 368–374 (1990)
56. Shunmugam, M.S., Bhaskara, R.S.V., Narendran, T.T.: Selection of optimal conditions in multi-pass face-milling using a genetic algorithm. Int. J. Mach. Tools Manuf. 40, 401–414 (2000)
57. Wang, Z.G., Rahman, M., Wong, Y.S., et al.: Optimization of multi-pass milling using genetic algorithm and genetic simulated annealing. Int. J. Mach. Tools Manuf. 45, 1726–1734 (2005)
58. Zarei, O., Fesanghary, M., Farshi, B., et al.: Optimization of multi-pass face-milling via harmony search algorithm. Journal of Materials Processing Technology (2008), doi:10.1016/j.jmatprotec.2008.05.029

Optimum Design of Steel Skeleton Structures

Mehmet Polat Saka

Engineering Sciences Department, Middle East Technical University, Ankara, Turkey
mpsaka@metu.edu.tr

Abstract. Designing steel skeleton structures optimally is a tedious task for designers. They need to select the appropriate steel sections for the members of a steel frame from a discrete list that is available in the practice. Assigning arbitrarily any one of these sections to any member of the steel frame generates one possible design combination. The total number of such combinations is quite large and finding out which one these combinations produces the optimum design by trial is almost practically impossible. The harmony search algorithm is a recently developed technique for determining the solution of such combinatorial optimization problems. In this chapter, harmony search algorithm based optimum design methods are introduced for moment resisting steel frames, grillage systems, geodesic domes and cellular beams. The design problems of these steel structures are carried out according to British and American Design Codes respectively. The design examples considered in both structural systems have shown that harmony search algorithm is an effective and robust technique to find the optimum solutions of combinatorial structural optimization problems.

Keywords: Structural Optimization, Optimum Design of Steel Structures, Metaheuristic Search Techniques, Harmony Search Method, Combinatorial Structural Optimization.

1 Introduction

Optimum structural design algorithms provide a useful tool to a steel designer by which she/he can determine the optimum topology, the optimum geometry and the optimum steel profiles for the members of a steel structure such that the steel structure can be constructed by using adequate steel material but not more. Structural design optimization achieves these goals such that design constraints specified by steel design codes are satisfied under the applied loads and the weight or the cost of the steel frame under consideration is the minimum. When formulated, two types of structural optimization problems can be distinguished. In some design problems the design variables can have continuous values. However in some others it is required that the values of design variables have to be selected from a set of discrete values. The optimum design problem of steel skeleton structures falls into the second category. A structural steel designer has to select steel sections from a discrete set which contains certain designations of steel profiles that are produced by steel mills. Hence the formulation of a steel frame design optimization problem turns out to be a discrete programming problem [1]. Obtaining solutions to discrete programming problems is more difficult than finding solutions to programming problems with continuous variables [2]. This may be one of the reasons why early mathematical programming techniques developed have dealt with continuous variables [3-5]. Later some of these algorithms have been extended to address discrete optimization problems as well. A comprehensive

Z.W. Geem (Ed.): Music-Inspired Harmony Search Algorithm, SCI 191, pp. 87–112.
springerlink.com

review of discrete structural optimization methods is given by Arora [6]. The algorithms that are based on mathematical programming techniques are deterministic. They need an initial design point to initiate a search for the optimum solution and require gradient computations in the exploration process.

Another group of optimization techniques that have emerged recently are non-deterministic approaches. These techniques do not require gradient information of the objective function and constraints, and use probabilistic transition rules not deterministic ones. They employ random number call, and incorporate a set of parameters that require to be adjusted prior to their use. These novel and innovative metaheuristic search algorithms make use of ideas inspired from the nature. The basic idea behind these techniques is to simulate natural phenomena, such as survival of the fittest, immune system, swarm intelligence and the cooling process of molten metals through annealing into a numerical algorithm [7-14]. The optimum structural design algorithms that are based on these techniques are robust and quite effective in finding solutions to discrete programming problems. A detailed review of these algorithms as well as their applications in optimum structural design is carried out by Saka [15]. One recent addition to these techniques is the harmony search algorithm [16-21]. This approach is based on the musical performance process that takes place when a musician searches for a better state of harmony. Jazz improvisation seeks to find musically pleasing harmony similar to the optimum design process which seeks to find optimum solution. The pitch of each musical instrument determines the aesthetic quality, just as the objective function value is determined by the set of values assigned to each decision variable. In this chapter optimum design algorithm based on harmony search method is presented for moment resisting steel frames, grillage systems and geodesic domes.

2 Harmony Search Method Based Optimum Design Algorithm

Consider a structural design problem where the total number of design variables is represented by *nv*. Let there be a design pool for each design variable that contains possible discrete values for that particular design variable. During the design process the optimum design algorithm is required to carry out a selection for a design variable among the values specified in the design pool of that particular design variable. In some design problems there may be only one design pool for all the design variables from which the selection is carried out, while in some other there can be more than one design pool depending on the type of design variables. Hence the total number of discrete values in design pools may vary depending on whether there is only one design pool for all the variables or not. Let us consider at this stage for simplicity that there is only one design pool from which the possible values of each design variable are to be selected. Let there be *ns* discrete values in this pool. Each of these values can be a possible candidate for the design variable during the design process. In the design problems considered in this chapter the sequence number in the design pool corresponding to each value is taken as design variable instead of the value itself. For example if a design pool has 64 discrete values for a design variable, an integer number between 1 to 64, say 24, specifies the particular value that can be adopted for the design variable under consideration. This provides ease and flexibility in the implementation of computer programming. Finding out the appropriate combination of these values such that the objective function holds its minimum value while design constraints are satisfied is a combinatorial optimization problem. Such combination constitutes the optimum solution to the design problem under consideration. As mentioned in previous section, harmony search method is found quite effective in obtaining the solution of such design problems. An optimum design algorithm that is based on harmony search method has the following steps:

Step 1. Initialization of Harmony Memory Matrix: A harmony memory matrix **H** is generated and initialized first. This matrix incorporates a specified number of solutions referred to as harmony size (*hms*). Each solution (harmony vector, $\mathbf{I}^i$) consists of *nv* integer number between 1 to *ns* selected randomly each of which corresponds sequence number of design variables in the design pool, and is represented in a separate row of the matrix; consequently the size of **H** is (*hms*×*nv*).

$$\mathbf{H} = \begin{bmatrix} I_1^1 & I_2^1 & \dots & I_{nv}^1 \\ I_1^2 & I_2^2 & \dots & I_{nv}^2 \\ \dots & \dots & \dots & \dots \\ I_1^{hms} & I_2^{hms} & \dots & I_{nv}^{hms} \end{bmatrix} \begin{matrix} \phi(\mathbf{I}^1) \\ \phi(\mathbf{I}^2) \\ \dots \\ \phi(\mathbf{I}^{hms}) \end{matrix} \quad (1)$$

I_i^j is the sequence number of the ith design variable in the jth randomly selected feasible solution.

Step 2. Evaluation of Harmony Memory Matrix: *hms* solutions shown in Eq. 1 are then analyzed, and their objective function values are calculated. The solutions evaluated are sorted in the matrix in the increasing order of objective function values, that is $\phi(\mathbf{I}^1) \leq \phi(\mathbf{I}^2) \leq \ldots \leq \phi(\mathbf{I}^{hms})$.

Step 3. Improvizing a New Harmony: A new harmony $\mathbf{I}' = [I_1', I_2', .., I_{nv}']$ is improvised (generated) by selecting each design variable from either harmony memory or the entire discrete set. The probability that a design variable is selected from the harmony memory is controlled by a parameter called harmony memory considering rate ($hmcr$). To execute this probability, a random number r_i is generated between 0 and 1 for each variable I_i. If r_i is smaller than or equal to $hmcr$, the variable is chosen from harmony memory in which case it is assigned any value from the *i*-th column of the **H**, representing the value set of variable in *hms* solutions of the matrix (Eq. 1). Otherwise (if $r_i > hmcr$), a random value is assigned to the variable from the entire discrete set.

$$I_i' = \begin{cases} I_i' \in \{I_i^1, I_i^2, ..., I_i^{\mu}\} & if \ r_i \leq hmcr \\ I_i' \in \{1, .., ns\} & if \ r_i > hmcr \end{cases} \quad (2)$$

If a design variable attains its value from harmony memory, it is checked whether this value should be pitch-adjusted or not. Pith adjustment simply means sampling the variable's one of the neighboring values, obtained by adding or subtracting one from its current value. Similar to $hmcr$ parameter, it is operated with a probability known as pitch adjustment rate (par) in Eq. 3. If not activated by par, the value of the variable does not change.

$$I_i'' = \begin{cases} I_i' \pm 1 & if\ r_i \le par \\ I_i' & if\ r_i > par \end{cases} \quad (3)$$

Constraint handling: The new harmony vector obtained using the above-mentioned rules is checked whether it violates design constraints. If this vector is severely infeasible it is discarded and another harmony vector is sought. However if it is slightly infeasible, it is included in the harmony matrix. In this way the slightly infeasible harmony vector is used as a base in the pitch adjustment operation to provide a new harmony vector that may be feasible. This is achieved by using large error values initially for the acceptability of the new design vectors. The error value is then gradually reduced during the design cycles until it reaches to its final value. This value is then kept the same until the end of iterations. This adaptive error strategy is found quite effective in handling the design constraints in large design problems.

Step 4. Update of Harmony Matrix: After generating the new harmony vector, its objective function value is calculated. If this value is better (lower) than that of the worst harmony vector in the harmony memory, it is then included in the matrix while the worst one is discarded out of the matrix. The updated harmony memory matrix is then sorted in ascending order of the objective function value.

Step 5. Termination: The steps 3 and 4 are repeated until a pre-assigned maximum number of cycles N_{cyc} is reached. This number is selected large enough such that within this number no further improvement is observed in the objective function.

3 Optimum Design of Steel Frames

Optimum design of steel frames necessitates determination of steel profiles from a standard steel sections table available in the practice for the beams and column of a frame such that its response under the applied load is within the limitations specified by steel design codes and its cost or weight is the minimum.

3.1 Discrete Optimum Design Problem

For a steel frame consisting of *nm* members that are collected in *ng* design groups (variables), the optimum design problem according to BS 5950 [22] code yields the following discrete programming problem, if the design groups are selected from steel sections in a given profile list.

Find a vector of integer values **I** (Eq. 1) representing the sequence numbers of steel sections assigned to *ng* member groups

$$\mathbf{I}^T = [I_1, I_2, \ldots, I_{ng}] \quad (4)$$

to minimize the weight (W) of the frame.

$$\text{Minimize } W = \sum_{r=1}^{ng} m_r \sum_{s=1}^{t} l_s \quad (5)$$

subject to

$$(\delta_j - \delta_{j-1})/h_j \le \delta_{ju}, \quad j = 1,\ldots, nsy \tag{6}$$

$$\delta_i \le \delta_{iu} \qquad i = 1,\ldots, nd \tag{7}$$

$$\frac{F_k}{A_{gk}\,Py} + \frac{M_{xk}}{M_{cxk}} \le 1 \tag{8}$$

$$\text{or} \quad \frac{F_k}{A_{gk}\,P_{ck}} + \frac{m_k M_{xk}}{M_{bk}} \le 1, \quad k = 1,\ldots, nc \tag{9}$$

$$M_{xn} \le M_{cxn}, \quad n = 1,\ldots, nb \tag{10}$$

where Eq. 5 defines the weight of the frame, m_r is the unit weight of the steel section selected from the standard steel sections table that is to be adopted for group r. t_r is the total number of members in group r and ng is the total number of groups in the frame. l_s is the length of member s which belongs to group r.

Eq. 6 represents the inter-storey drift of the multi-storey frame. δ_j and δ_{j-1} are lateral deflections of two adjacent storey levels and h_j is the storey height. nsy is the total number of storeys in the frame. Eq. 7 defines the displacement restrictions that may be required to include other than drift constraints such as deflections in beams. nd is the total number of restricted displacements in the frame. δ_{ju} is the allowable lateral displacement. BS 5950 limits the horizontal deflection of columns due to unfactored imposed load and wind loads to height of column / 300 in each storey of a building with more than one storey. δ_{iu} is the upper bound on the deflection of beams which is given as span / 360 if they carry plaster or other brittle finish.

Eq. 8 defines the local capacity check for beam-columns. F_k and M_{xk} are respectively the applied axial load and moment about the major axis at the critical region of member k. A_{gk} is the gross cross sectional area, and p_y is the design strength of the steel. M_{cx} is the moment capacity about major axis. nc is the total number of beam-columns in the frame. Eq. 9 represents the simplified approach for the overall buckling check for beam-columns. m is the equivalent uniform moment factor given in BS 5950. M_{bk} is the buckling resistance moment capacity for member k about its major axis computed from clause 4.3.7 of the code. p_{ck} is the compression strength obtained from the solution of quadratic Perry-Robertson formula given in appendix C.1 of BS 5950. It is apparent that computation of compressive strength of a compression member requires its effective length. This can be automated by using Jackson and Moreland monograph for frame buckling [23]. The relationship for the effective length of a column in a swaying frame is given as:

$$\frac{(\gamma_1\gamma_2)(\pi/k)^2-36}{6(\gamma_1+\gamma_2)}=\frac{\pi/k}{\tan(\pi/k)} \tag{11}$$

where k is the effective length factor and γ_1 and γ_2 are the relative stiffness ratio for the compression member which are given as:

$$\gamma_1=\frac{\sum I_{c1}/\ell_{c1}}{\sum I_{b1}/\ell_{b1}},\quad \gamma_2=\frac{\sum I_{c2}/\ell_{c2}}{\sum I_{b2}/\ell_{b2}} \tag{12}$$

The subscripts c and b refer to the compressed and restraining members respectively and the subscripts 1 and 2 refer to two ends of the compression member under investigation. The solution of the nonlinear equation (Eq. 11) for k results in the effective length factor for the member under consideration. Eq. 11 has the following form for non-swaying frames.

$$\frac{\gamma_1\gamma_2}{4}\left(\frac{\pi}{k}\right)^2+\left(\frac{\gamma_1+\gamma_2}{2}\right)\left(1-\frac{\pi/k}{\tan(\pi/k)}\right)+\frac{2\tan(\pi/2k)}{\pi/k}=1 \tag{13}$$

Eq. 10 defines the moment capacity check for beams. The details of the computation of these are given in [24]. Further to the above restrictions the geometric constraints that are necessary for the constructability of a steel frame are also considered. These geometric constraints are such that the beam section at each beam-column connection at each storey should be less than or equal to the flange width of column section and the depth and the mass per meter of column section at each storey and at each beam-column connection should be less than or equal to width and mass of the column section at the lower storey.

3.2 Design Example

Three-bay, fifteen-storey frame shown in Figure 1 is designed by optimum design algorithm described in section 2. The dimensions of the frame and the loading are shown in the figure. The frame is subjected to gravity loading of 12.4kN/m on the beams of roof level and 27kN/m on the beams of each floor. The lateral loading is the wind loading that is computed according to British Code considering 45m/s wind speed and 6m frame spacing. The modulus of elasticity is 200kN/mm^2. Frame consists of 105 members that are collected in 12 groups. Columns in every story made out of the same section. The beams of roof and intermediate floors are considered to be two different groups as shown in the figure. The allowable inter-storey drift is 11.7mm while the lateral displacement of the top storey is limited to 176.7mm. The strength capacities of steel members are computed according to BS5950.

Two design pools, one for the beams of the frame and the other for the columns of the frame are established before initiating the search for the optimum design. Among the steel sections list of BS 5950 [25], 64 Universal Beam (UB) sections starting from 914×419×388 to 254×102×28 and 32 Universal Column (UC) sections starting from 356×406×634 to 152×152×23 are selected to constitute two discrete sets which are used as design pools.

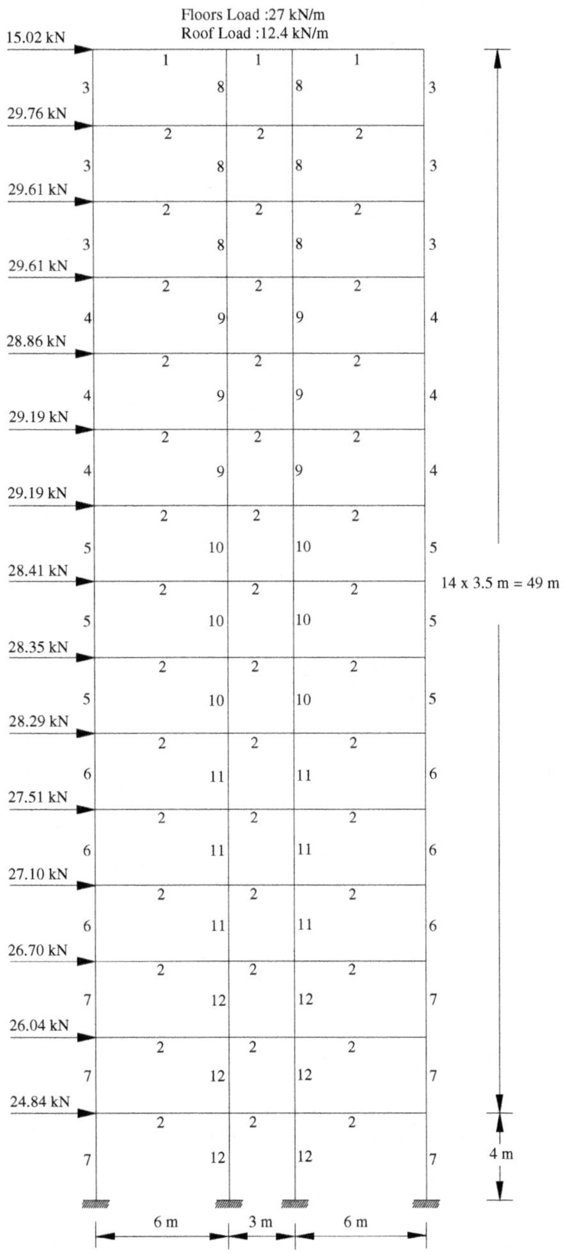

Fig. 1. Fifteen-storey, three-bay moment resisting steel frame

The optimum steel profiles obtained by the harmony search algorithm are given in Table 1 (UB Sections are for beams and UC sections are for columns). This design required 25,000 frame analysis 20,000 of which is needed for the improvisation of the

Table 1. Optimum designs for fifteen-storey, three-bay moment resisting steel frame

Group No	Harmony Search Algorithm	Genetic Algorithm
1	406×140×39 UB	305×102×25 UB
2	457×191×67 UB	457×152×82 UB
3	203×203×46 UC	254×254×73 UC
4	203×203×60 UC	254×254×89 UC
5	254×254×73 UC	254×254×107 UC
6	305×305×97 UC	356×368×129 UC
7	356×368×129 UC	356×368×129 UC
8	203×203×52 UC	203×203×46 UC
9	305×305×97 UC	254×254×73 UC
10	356×368×129 UC	356×368×129 UC
11	356×368×153 UC	356×368×153 UC
12	356×368×153 UC	356×368×202 UC
Minimum Weight (kg)	35723	41676

harmony memory matrix. It is noticed that harmony search algorithm has to go through large amount of trials to determine a feasible or near feasible solutions that satisfy the geometric constraints. The minimum weight of the frame is 35,723kg. It is observed that in the optimum frame the drift constraint for the tenth floor was at its upper bound of 11.7mm while the lateral displacement of top storey was 119.4mm against its upper bound of 176.7mm. The highest ratio among the combined strength constraints was 0.77 compare to 1 which was attained in the inner column of tenth storey. This clearly indicates that drift constraints under the wind loading dominates the design. The same frame was also designed using the genetic algorithm under the same design specifications in [26] and the optimum design achieved in that study is 41,676kg. The design obtained by the harmony search algorithm is 14% lighter than the one obtained by the simple genetic algorithm.

4 Optimum Design of Grillage Systems

Grillage system is a plane structure as shown in Figure 2 composed of longitudinal and transverse beams that intersect one another. The connections at intersection points between the members are assumed to be rigid and grillage systems carry loads that are applied out of their plane. They are widely utilized in bridge decks, ship hulls and building floors. Optimum design of grillage systems aims at finding the cross sectional properties of transverse and longitudinal beams such that the response of the system under the external loading is within the allowable limits defined in a code of practice while the weight or the cost of the system is the minimum.

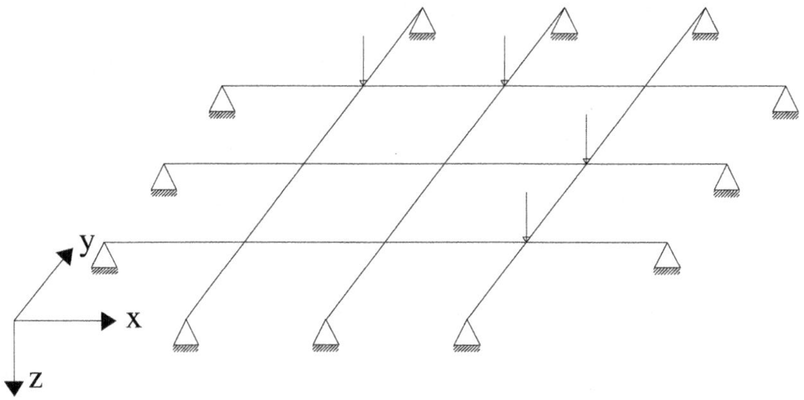

Fig. 2. Steel Grillage system

4.1 Discrete Optimum Design Problem

The optimum design problem of a typical grillage system shown in Figure 2 where the behavioral and performance limitations are implemented from LRFD-AISC (Load and Resistance Factor Design - American Institute of Steel Construction) [27] can be expressed as in the following.

Find a vector of integer values **I** (Eq. 14) representing the sequence numbers of W-sections given in W-section list of LRFD-AISC assigned to *ng* member groups

$$\mathbf{I}^T = \left[I_1, I_2, \ldots, I_{ng}\right] \tag{14}$$

to minimize the weight (W) of the grillage system.

$$\text{Minimize} \quad W = \sum_{k=1}^{ng} m_k \sum_{i=1}^{n_k} l_i \tag{15}$$

subject to

$$\delta_j / \delta_{ju} \leq 1, \quad j = 1,2,\ldots, p \tag{16}$$

$$M_{ur} / (\phi_b M_{nr}) \leq 1, \quad r = 1,2,\ldots, nm \tag{17}$$

$$V_{ur} / (\phi_v V_{nr}) \leq 1, \quad r = 1,2,\ldots, nm \tag{18}$$

where m_k in Eq. 15 is the unit weight of grillage element belonging to group k to be selected from W-sections list of LRFD-AISC, n_k is the total number of members in group k, and ng is the total number of groups in the grillage system. l_i is the length of member i . δ_j in Eq. 16 is the displacement of joint j and δ_{ju} is its upper bound. The joint displacements are computed using the matrix displacement method for grillage systems. Eq. 17 represents the strength requirement for laterally supported beam in load and resistance factor design according to LRFD-F2. In this inequality ϕ_b is the resistance factor for flexure which is given as 0.9, M_{nr} is the nominal moment strength and M_{ur} is the factored service load moment for member *r*.

Eq. 18 represents the shear strength requirement in the load and resistance factor design according to LRFD-F2. In this inequality ϕ_v represents the resistance factor for shear given as 0.9, V_{nr} is the nominal strength in shear and V_{ur} is the factored service load shear for member *r*. The details of obtaining nominal moment strength and nominal shear strength of a W-section according to LRFD are given in [27].

4.2 Design Example

The grillage system shown in Figure 3 is used to cover the area of 12m×12m. The grillage system is planned to carry 15kN/m^2 uniformly distributed load covering the whole area total of which is 2,160kN. This total load can be applied to the system as 86.4kN of point load acting at each joint. The grillage system has 60 members altogether that are collected in two groups. The vertical displacements of joints 8, 13, 14 and 18 are restricted to 25 mm. A36 mild steel is selected for the design which has the yield stress of 250MPa, the modulus of elasticity of 205kN/mm^2 and shear modulus of 81kN/mm^2 respectively. The discrete set from which the design algorithm selects the sectional designations for grillage members is considered to be the complete set of

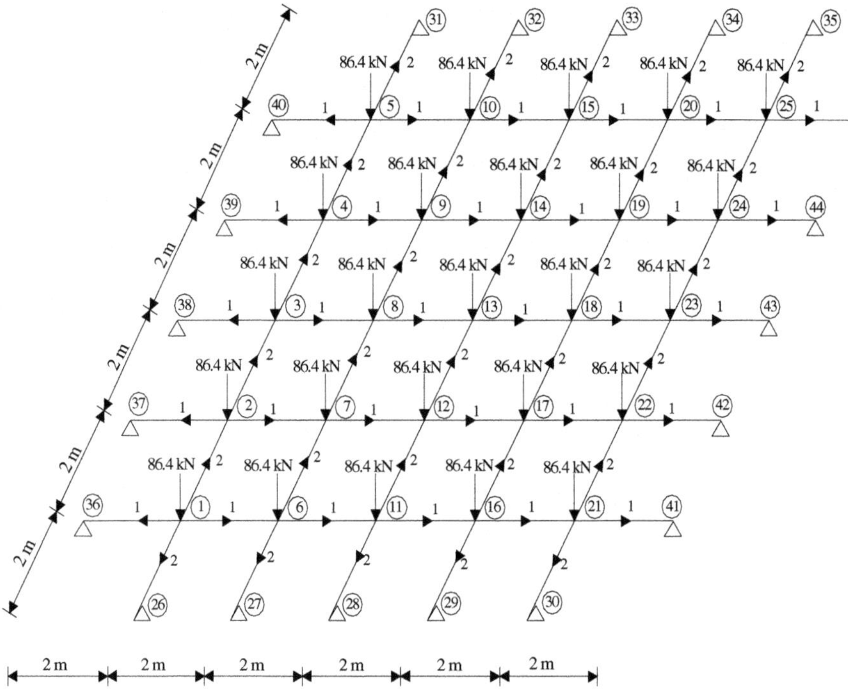

Fig. 3. 60-member grillage system

272 W-sections starting from W100×19.3 to W1100×499mm as given in LRFD-AISC [27]. The sequence number of each section in the set is used as design variable. Hence the terms of the harmony memory matrix represents the sequence number of W-sections in the discrete set.

After carrying out a sensitivity analysis with different values of harmony search parameter, it is noticed that the values of 50 for *hms*, 0.9 for *hmcr* and 0.5 for *par* respectively produces the grillage with the minimum weight. With these parameters the optimum design obtained for the system is given in Table 2. Harmony search algorithm has selected W200×22.5 section for the first group and W690×217 section for the second group in the optimum design which has the minimum weight of 14,384kg [28]. Furthermore it is observed that the vertical displacement constraint is at its upper bound, the strength ratio is only 0.48 which is much less than 1. This clearly reveals the fact that the serviceability constraints are dominant in the design problem. The optimum result presented in Table 2 is obtained after 10,000 iterations of the harmony search method. However, it was noticed that the optimum sectional designations remained the same after 4,000 iterations.

Table 2. Optimum Design for 60-member grillage system

Optimum W-Section Designations		δ_{max} (mm)	Maximum Strength Ratio	Minimum Weight (kg)
Group 1	Group 2			
W200×22.5	W690×217	25.0	0.48	14,384

The harmony search method based optimum design algorithm presented above is also used to demonstrate the effect of beam spacing in the optimum design of grillages [29]. For this purpose, it is decided that the square area of 12m×12m is to be covered with grillage systems each having different beam spacing. The grillage system that can be used to cover the area will have 12m long the longitudinal beams and 12m long the transverse beams. For this purpose five different grillage systems are considered having 3m, 2.4m, 2m, 1.5m and 1m beam spacing. The total external load of 2,160kN is distributed to joints of the grillage system as a point load value of which is calculated according to the beam spacing. A36 mild steel is selected for the design which has the yield stress of 250MPa, the modulus of elasticity of 205kN/mm^2 and shear modulus of 81kN/mm^2 respectively. The members of grillage structures are collected in two groups. The longitudinal beams are considered to belong to group 1 and transverse beams are taken as group 2. Harmony search parameters; Harmony memory size (*hms*) is taken as 50, Harmony memory considering rate (*hmcr*) is selected as 0.9 while pitch adjusting rate (*par*) is considered as 0.5 on the basis of sensitivity analysis.

Each grillage system with five different spacing is designed using the design algorithm. It is noticed that the minimum weight of grillage systems changes with the beam spacing. There is a decrease from 3-m beam spacing to 2-m beam spacing, then the minimum weight increases. The variation of the minimum weight with the beam spacing is shown in Figure 4. In this study, the beam spacing is selected as numbers

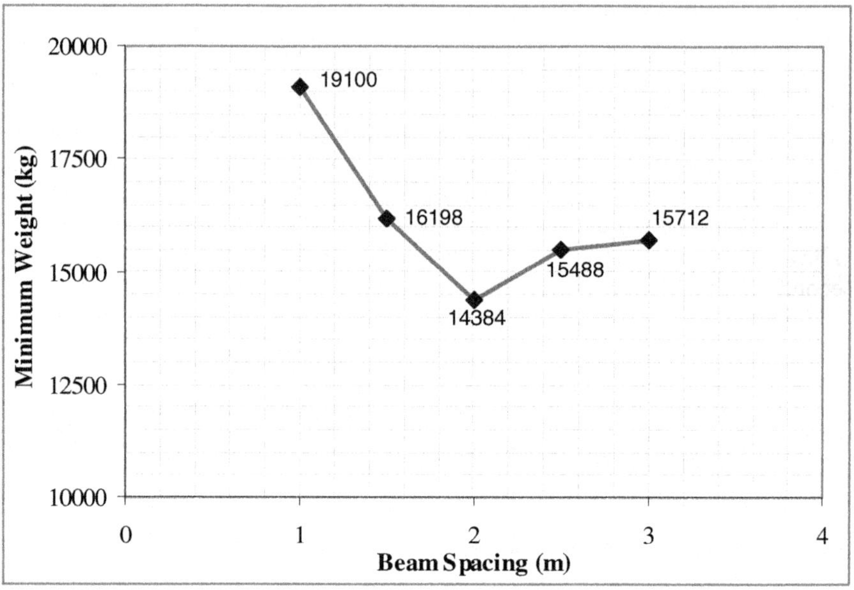

Fig. 4. Variation of weight different beam spacing

that are practically preferred. It is apparent from Figure 4 that 2-m spacing is the optimum spacing among the values considered. Consequently it should be pointed out that consideration of beam spacing as an additional design variable in addition to steel section designations that are selected for member groups in the design of grillage systems would yield better results.

5 Optimum Geometry Design of Geodesic Domes

Domes are lightweight and cost effective structures that are used to cover large areas such as exhibition halls, stadium and concert halls. They provide a completely unobstructed inner space and they are economical in terms of materials compared to the more conventional forms of structures as explained in Makowski [30]. They consist of one or more layers of elements that are arched in all directions. These structures are sometimes called braced domes which is a term used for single-layer spherical space structures that are usually used to cover large spans up to about 150m and have typically a very small weight, around 15-20 kg/m^2.

The geodesic dome shown in Fig. 5 is a commonly used form as a structural system. It has relatively simple geometry. In this dome all the structural data related with the geometry of the dome can be obtained automatically provided that the diameter D, the total number of rings n_r and the height of the crown h in the dome are known. It is worthwhile to mention that rings in the dome are arranged in such a way that the distance between them on the meridian line is equal. The details of this automatic data generation are given in [31].

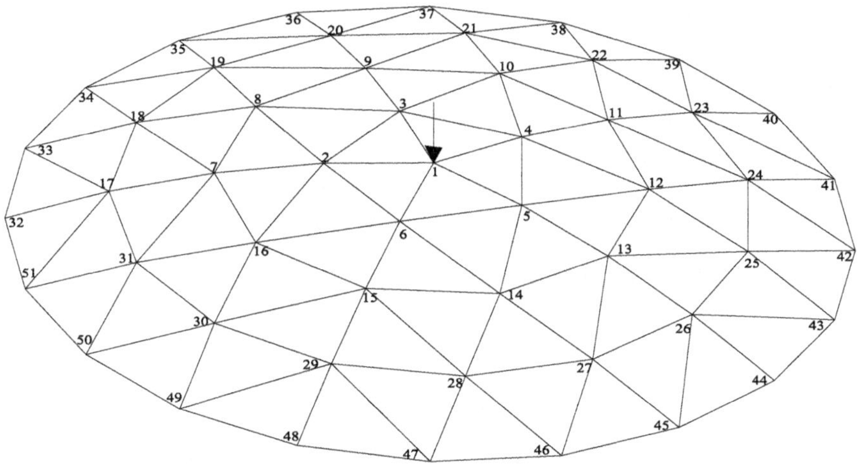

Fig. 5. Geodesic dome

5.1 Optimum Geometry Design Problem

The optimum geometry design of a geodesic dome requires the selection of steel sections such as circular hollow sections or other types for its members from a standard steel sections table and the height of the dome. This selection should be carried out such a way that the dome obtained satisfies the serviceability and strength requirements specified by the code of practice while the economy is observed in the overall or material cost of the structure.

The discrete optimum geometry design problem of geodesic dome where the minimum weight is taken as the objective and the optimization constraints mentioned above are implemented from BS 5950 [22], has the following form:

Find a vector of integer values **I** (Eq. 19) representing the sequence numbers for total of *nv* design variables that consist of the discrete values given in the set arranged for the height of dome and sequence number of circular hollow section given in BS5950 that are to be assigned to *ng* member groups

$$\mathbf{I}^T = [I_1, I_2, \ldots, I_{nv}] \tag{19}$$

to minimize the weight (W) of the geodesic dome.

$$\text{Minimize} \quad W = \sum_{r=1}^{ng} m_r \sum_{s=1}^{t_r} l_s \tag{20}$$

subject to

$$\delta_i \le \delta_{iu}, \quad i = 1,2,\ldots\ldots, nd \tag{21}$$

$$\frac{F_{ck}}{Ag_k P_y} + \frac{M_{xk}}{M_{cxk}} + \frac{M_{yk}}{M_{cyk}} \le 1, \quad k = 1,2,\ldots, nm \tag{22}$$

$$\frac{F_{ck}}{P_{ck}} + \frac{m_{LTK} M_{LTk}}{M_{bk}} + \frac{m_y M_{yk}}{P_y Z_{yk}} \leq 1 \tag{23}$$

Eq. 20 defines the weight of the dome where the minor weight of connections is ignored. The weight is function of the unit weight m_r of the steel section selected for group r in the dome from the standard steel section table. t_r is the total number of members in group r and ng is the total of groups in the dome.

Eq. 21 defines the displacement limitations that are required for the serviceability requirements. BS5950 limits the vertical deflections under the unfactored imposed load to span/360 and horizontal displacements to the height/300. nd is the total number of restricted nodal displacements in the dome. δ_i represents one of such displacement and δ_{iu} is its upper bound.

Eqs. 22 and 23 define the cross-sectional capacity and the member buckling resistance checks for compression members with bi-axial moments. These expressions are given in clauses 4.8.3. and 4.8.3.3 of BS 5950. They ensure that at the points of the greatest bending moment and axial load, yielding or local buckling does not take place. In Eq. 22, F_{ck} is the axial compression, M_{xk} is the maximum major axis moment in the segment length ℓ_x governing P_{cx} and M_{yk} is the maximum minor axis moment in segment length ℓ_y governing P_{cy}. M_{cxk} and M_{cyk} are the moment capacity of member k about its major and minor axis. A_{gk} is the gross cross sectional area and p_y is the design strength of the steel grade used for member k respectively. Eq. 23 is the simplified approach for checking the member buckling resistance. P_{ck} is the compression resistance of the member k, which is the smaller of P_{cx} and P_{cy}. These are the compression resistance of the member considering buckling about the major axis and the minor axis respectively. The compression resistance of a member is equal to the gross cross sectional area times the compression strength p_{cy} which is obtained from the solution of the Perry-Robertson quadratic equation given appendix C.1. of BS 5950. It is apparent that computation of the compression strength of a member requires its effective length. For simplicity the effective lengths of members are taken as their lengths. m_{LT} is the equivalent uniform moment factor given in the code. M_{LT} is the maximum major axis moment in the segment length L governing the buckling resistance moment M_b as explained in clause 4.3.6 of BS5950. For circular hollow steel sections M_b is equal to moment capacity M_c of the section. Z_{yk} is the section modulus about the minor axis. nm is the total number of members in the dome.

5.2 Design Example

The design algorithm presented is used to determine the optimum height and circular steel hollow section designations for the geodesic dome shown in Figure 5. The modulus of elasticity is taken as 205kN/mm^2. The grade of steel adopted is Grade 43.

The dome is considered to be subjected to equipment loading of 1000kN at its crown. The limitations imposed on the joint displacements as per BS5950. X and Y displacements of joints 1, 2 and 3 are restricted to 33mm and Z displacements of the same joints are limited to 28mm. Among the circular hollow sections given in [24] 64 sections that vary from PIP213.2 to PIP1936.3 are selected to be used as a design pool for the harmony search algorithm to select from. For the height of the crown a list is prepared which starts from 1m to 8.75m with the increment of 0.25m. There are 32 values altogether for the harmony search algorithm to choose from.

The member grouping in a design of dome is decided by the designer. In the example under consideration, it is decided that those members between each ring are to be made one group and the members on each ring are another group. For example if the grouping is carried out such a way that the diagonal members between the crown and the first ring are group 1, the members on the first ring are group 2, the members between ring 1 and 2 are group 3 and the group number of members on the ring 2 is 4 and so forth, then the total number of groups in the dome becomes twice the total number of rings in the design problem. Hence in the case of three rings in the dome the total number of design variables is 7, six of which are the sectional designation to be selected for each group and the last one is the height of the dome. In the case of six rings in the dome the total number of design variables is 13, twelve of which represents the sectional designations to be adopted for each group and the last one represents the crown height in the dome.

Harmony memory matrix size (*hms*), harmony memory considering rate (*hmcr*) and pitch-adjusting rate (*par*) are taken as 30, 0.90 and 0.45 after carrying out a sensitivity analysis with various values of these parameters. The maximum number of iterations is selected as 10,000 after number of different trials.

The optimum sectional designations for each group obtained for the dome with three rings by the presented algorithm are given in Table 3. The optimum height of the dome is determined as 2m. The maximum displacement in the dome is 31.7mm and the maximum strength ratio is 1.00. The minimum weight of the dome is attained as 1,244.42kg. The optimum geometry of the dome obtained is shown in Figure 6(a).

Table 3. Optimum Pipe Sections for Geodesic Dome with Three Rings

Optimum Pipe Section Designations for Member Groups					
1	2	3	4	5	6
PIP 1,393.6	PIP 1,143.0	PIP 603.6	PIP 483.2	PIP 423.2	PIP 213.2

The optimum sectional designations for each group obtained for the dome with four rings are given in Table 8.4. The optimum height of the dome is determined as 2m. The maximum displacement in the dome is 30.4mm and the maximum strength ratio is 1.00. The minimum weight of the dome is attained as 1335.14kg.The optimum geometry of the dome obtained in is shown in Figure 6(b).

Table 4. Optimum Pipe sections for Geodesic Dome with Four Rings

Optimum Pipe Section Designations for Member Groups							
1	2	3	4	5	6	7	8
PIP 1393.6	PIP 1143.0	PIP 485.0	PIP 424.0	PIP 423.0	PIP 333.0	PIP 333.2	PIP 213.2

The optimum sectional designations for each group obtained for the dome with five rings are given in Table 5. The optimum height of the dome is determined as 1.5m. The maximum displacement in the dome is 32mm and the maximum strength ratio is 0.99. The minimum weight of the dome is attained as 1,477.08kg. The optimum geometry of the dome obtained is shown in Figure 6(c).

Table 5. Optimum Pipe sections for Geodesic Dome with Five Rings

Optimum Pipe Section Designations for Member Groups									
1	2	3	4	5	6	7	8	9	10
PIP 886.3	PIP 765.0	PIP 763.6	PIP 483.6	PIP 423.2	PIP 333.2	PIP 333.0	PIP 422.6	PIP 263.2	PIP 213.2

The optimum sectional designations for each group obtained for the dome with six rings are given in Table 6. The optimum height of the dome is determined as 1.5m. The maximum displacement in the dome is 29.3mm and the maximum strength ratio is 1.00. The minimum weight of the dome is attained as 1545.3kg. The optimum geometry of the dome obtained is shown in Figure 6(d).

Table 6. Optimum Pipe sections for Geodesic Dome with Six Rings

Optimum Pipe Section Designations for member groups											
1	2	3	4	5	6	7	8	9	10	11	12
PIP 886.3	PIP 884.0	PIP 763.6	PIP 483.6	PIP 483.6	PIP 333.6	PIP 333.0	PIP 263.2	PIP 263.2	PIP 263.2	PIP 263.2	PIP 213.2

The maximum displacement values in each design problem are less than but close to their upper bound of 33mm while the maximum value of strength constraints are at their upper value of 1. This indicates that the displacement limitations are not dominant in these design problems. Hence the strength constraints are the ones that decide the optimum shapes in the design example considered. Variation of the minimum weights of geodesic domes with different number of rings is shown in Figure 7.

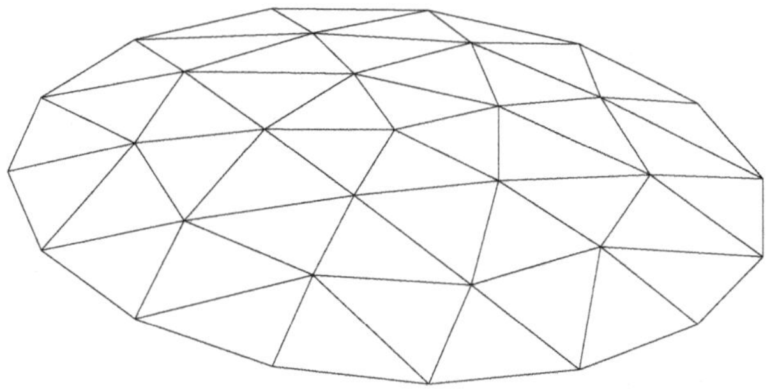

(a) Optimum geometry for geodesic dome with three rings

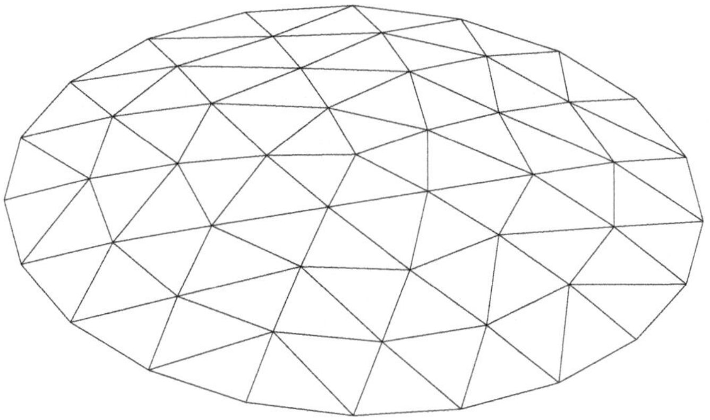

(b) Optimum geometry for geodesic dome with four rings

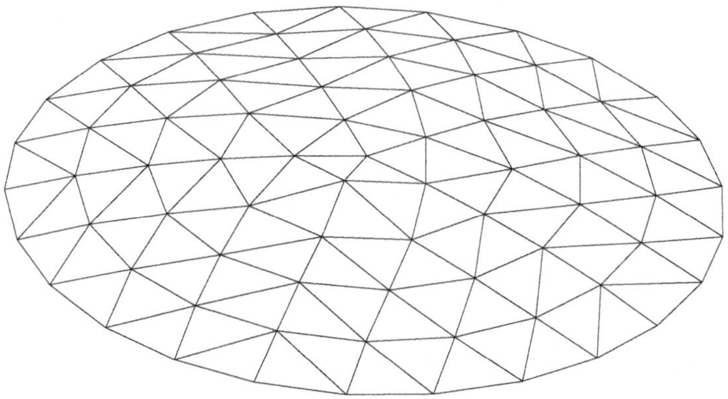

(c) Optimum geometry for geodesic dome with five rings

Fig. 6. Optimum heights for geodesic domes with different number of rings

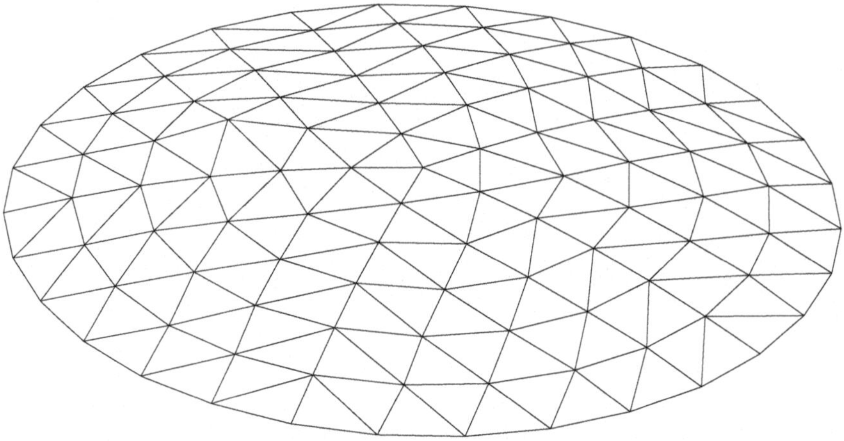

(d) Optimum geometry for geodesic dome with six rings

Fig. 6. (*continued*)

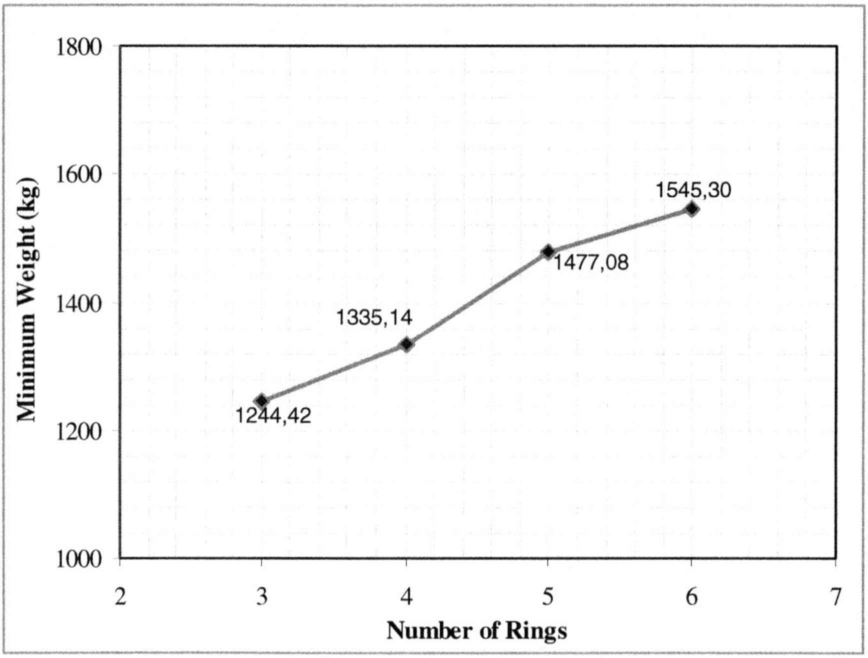

Fig. 7. Variation of the minimum weights of geodesic domes with different number of rings

6 Optimum Design of Cellular Beams

Common steel I-beam sections can be modified to intensify their strength by creating an open-web section from a root beam. This is achieved by cutting the web of the root beam in a certain pattern and then re-welding the two parts to each other. As a result of this process the overall beam depth increases that causes increase in the capacity of section. Cellular beams are steel sections with circular openings that are made by cutting a rolled beams web in a half circular pattern along its centerline and re-welding the two halves of hot rolled steel sections as shown in Figure 8. This circular opening up of the original rolled beam increases the overall beam depth, moment of inertia and section modulus, while reducing the overall weight of the beam. Cellular beams have been used in over 3,500 projects in over twenty countries [32]. The most common building types for the cellular beams are office buildings, parking garages, shopping centers and any structure with a suspended floor. These beams are approximately 40-60% deeper and 40-60% stronger than the original member while reducing the overall weight.

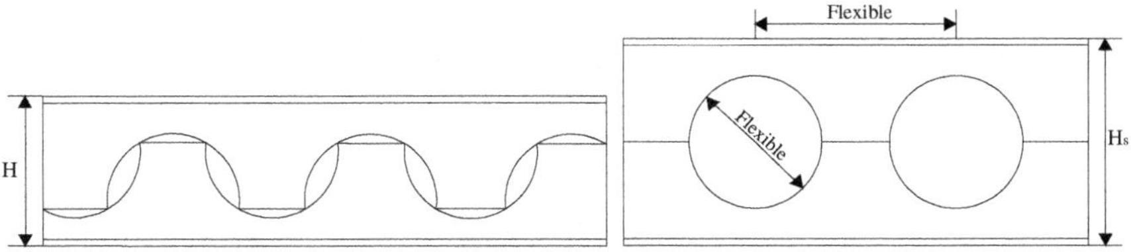

Fig. 8. Cellular beam

6.1 Optimum Design Problem

The design of a cellular beam requires the selection of a rolled beam from which the cellular beam is to be produced, the selection of circular hole diameter and the selection of spacing between the centres of these circular holes or total number of holes in the beam as shown in Figure 9. Hence the sequence number of the rolled beam sections in the standard steel sections tables, the circular hole diameter and the total number of holes are taken as design variables in the optimum design problem considered. For this purpose a design pool is prepared which consists of list of standard rolled beam sections, a list of various diameter sizes and a list of integer numbers starting from 2 to 40 for the total number of holes in a cellular beam. The optimum design problem formulated by considering the design constraints explained in The Steel Construction Institute Publication titled "Design of Composite and Non-composite Cellular Beams" [33] which are consistent with BS5950 [25]; part 1 and 3 yields the following mathematical model.

Find a integer design vector $\{I\}=\{I_1,I_2,I_3\}^T$ where I_1 is the sequence number of the rolled steel profile in the standard steel sections list, I_2 is the sequence number for the hole diameter in the discrete set which contains various diameter values and I_3 is the total number of holes for the cellular beam. Once I_1 is selected, then the rolled steel beam designation becomes known and all cross sectional properties of the

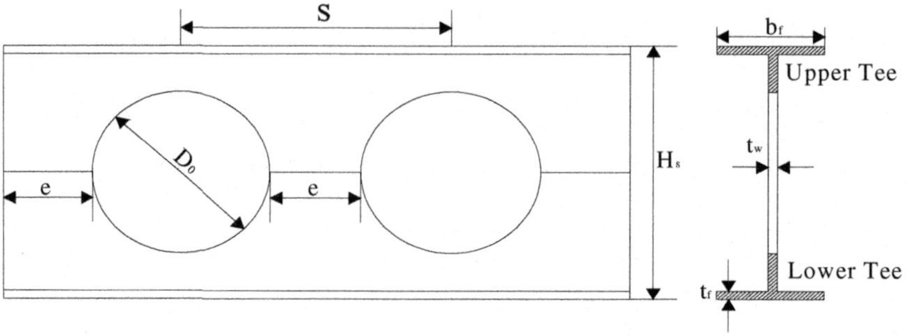

Fig. 9. Design variables for a cellular beam

beam becomes available for design. The corresponding values to I_2 and I_3 in the design sets makes the hole diameter and the total number of holes available for the cellular beam. Hence the design problem turns out to be as follows:

$$\text{Maximize} \quad W = \rho A L - \left(\pi \left(\frac{D_0}{2} \right)^2 NH \right) \tag{24}$$

subject to

$$g_1 = 1.08 \times D_0 - S \leq 0 \tag{25}$$

$$g_2 = S - 1.6 \times D_0 \leq 0 \tag{26}$$

$$g_3 = 1.25 \times D_0 - H_S \leq 0 \tag{27}$$

$$g_4 = H_S - 1.75 \times D_0 \leq 0 \tag{28}$$

$$g_5 = M_U - M_P \leq 0 \tag{29}$$

$$g_6 = V_{\max\sup} - P_v \leq 0 \tag{30}$$

$$g_7 = V_{O\max} - P_{vy} \leq 0 \tag{31}$$

$$g_8 = V_{H\max} - P_{vh} \leq 0 \tag{32}$$

$$g_9 = M_{A-A\max} - M_{w\max} \leq 0 \tag{33}$$

$$g_{10} = V_{Tee} - 0.5 \times P_{vy} \leq 0 \tag{34}$$

$$g_{11} = \frac{P_0}{P_u} - \frac{M}{M_p} - 1 \leq 0 \tag{35}$$

$$g_{12} = y_{\max} - L/360 \leq 0 \tag{36}$$

where W is the weight of the cellular beam, D_0 is hole diameter, ρ is density of steel, A is total area of the profile selected, NH is number of holes, H_S and L are the overall depth and the span of the cellular beam and S is the distance between centers of holes. M_U is the maximum moment under the applied loading, M_P is plastic moment capacity, $V_{\max\,\sup}$ is the maximum shear at support, $V_{O\max}$ is the maximum shear at the opening, $V_{H\max}$ is the maximum horizontal shear, $M_{A-A\max}$ is the maximum moment at A-A section shown in Figure 10, $M_{w\max}$ is the maximum allowable web post moment, V_{Tee} is vertical shear on tee, P_0, M forces on the section and $y_{\max}$ maximum deflection of the cellular beam.

Although the diameter of holes and spacing between their centers are left to designer to select, this selection has to satisfy the limitations given in constraints Eqs. 25-28. Inequality (Eq. 29) represents overall beam flexural capacity constraint. Under unfavourable applied load combinations the cellular beam should have sufficient flexural capacity to be able to resist the external loading. $M_p = A_{tee} p_y h$ where A_{tee} is the area of lower tee, p_y is the design strength of steel and h is the distance between centrals of upper tee and lower tee.

Inequalities (Eqs. 30-32) represent shear capacity checks. There are three shear checks in the design of cellular beams. The first one is the shear check at the support. Constraint (Eq. 30) makes sure that shear at the support does not exceed the shear capacity of the section where $P_v = 0.6 \times p_y \times$ ($0.9\times$*Area of web at supports*). It is also necessary to check two more shear failure modes additionally. The first shear failure mode check (Eq. 31) is the vertical shear capacity check of the beam. The sum of the shear capacities of the upper and lower tees gives the vertical shear capacity of the beam. The factored shear force in the beam should not exceed $P_{vy} = 0.6 \times p_y \times$($0.9 \times$ *Area of web at upper and lower tees*). The other (Eq. 32) is the horizontal shear check. The horizontal shear is developed in the web post due the change in axial forces in the tee as shown in Figure 10. The horizontal shear capacity in the web post of the beam should not exceed $P_{vh} = 0.6 \times p_y \times$($0.9 \times$ *Minimum area of web post*). The details of the computations of shear force and bending moment at a section of cellular beam is given in [34].

The flexural capacity of the upper and lower tees under bending is critical in cellular beams. The transfer of shear forces across a single opening causes secondary bending stresses. Inequalities (Eqs. 33-35) are required for the flexural and buckling strength of web post. The details of the computation of the maximum moment $M_{A-A\max}$ at section A-A shown in Figure 10 and the maximum allowable web post moment $M_{w\max}$ are given in [34]. The last constraint is the serviceability requirement that the cellular beam has to satisfy. This constraint requires that the maximum deflection of the beam should not be more than its span over 360.

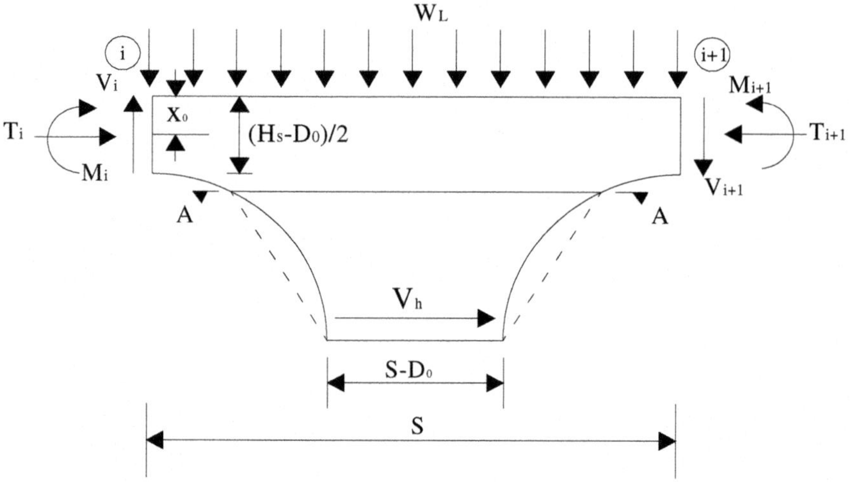

Fig. 10. Horizontal and vertical shear forces in web post of a cellular beam

6.2 Design Example

Harmony search method based optimum design algorithm presented in the previous sections is used to design a simply supported cellular beam with a span of 9m. The beam carries a uniform load of 40kN/m including its own weight and two concentrated loads of 50kN as shown in Figure 11. The allowable displacement of the beam is limited to 25 mm. The modulus of elasticity is taken as 205kN/mm^2 and Grade 50 is selected for the steel of the beam which has the design strength of 355MPa. Among the steel sections list of British Standards [25] 64 Universal Beam (UB) sections starting from 254×102×28 UB to 914×419×388 UB are selected to constitute the discrete set for steel sections from which the design algorithm selects the sectional designations for the cellular beam. In the design pool of hole diameters 421 values are arranged which varies between 180 and 600mm with increment of 1mm. The discrete set for the number of holes starts with 2 and goes up to 40 with increment of 1.

The optimum design of the cellular beam described is carried out by the design algorithm presented. The algorithm necessitates selection of harmony search parameters

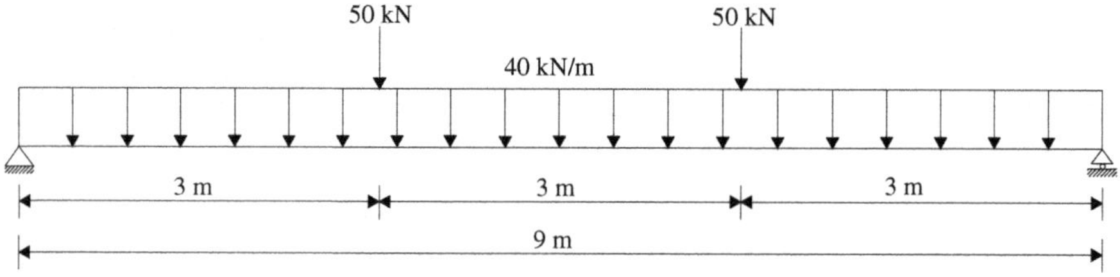

Fig. 11. Simply supported beam

Table 7. Trial of various harmony search algorithm parameters

Cases	*Hms*	*hmcr*	*par*	Weight (kg)
1	10	0.70	0.25	1,028.79
2	30	0.85	0.35	1,017.11
3	10	0.75	0.50	1,014.40
4	20	0.90	0.20	971.86
5	20	0.80	0.45	968.16
6	30	0.90	0.30	937.32
7	10	0.75	0.45	933.41
8	20	0.85	0.40	931.87
9	30	0.80	0.25	929.59
10	30	0.70	0.30	925.24

Table 8. Optimum design of cellular beam with 9m span

Optimum UB-Section Designations	Hole Diameter (mm)	Total Number of Holes	Maximum Strength Ratio	Minimum Weight (kg)
533×210×109 UB	586	11	0.98	925.24

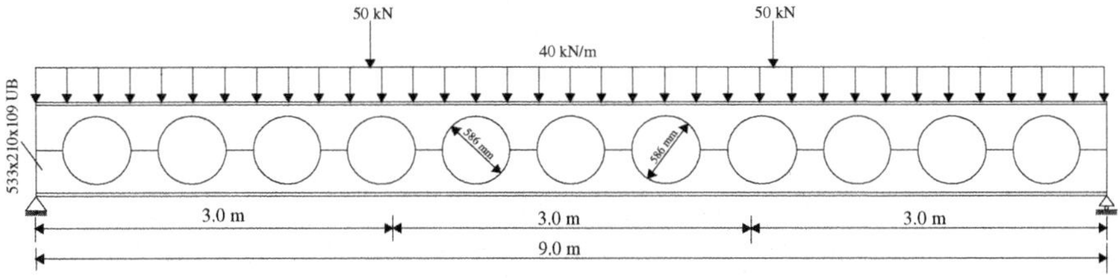

Fig. 12. Optimum profile of the cellular beam

prior to its use. For this reason a parametric study is carried out with different values of these parameters and the results obtained is shown in Table 7.

The table covers 10 different cases and for each case the total number of iterations is taken as 10,000. It is apparent from the table that the minimum weight for the cellular beam is obtained with *hms*=30, *hmcr*=0.70 and *par*=0.30. In the optimum design harmony search algorithm selects 533x210x109 UB profile, 11 circular holes and 586 mm for the hole diameter. This design has the minimum weight of 925.24 kg. The optimum result is given in Table 8. The maximum strength ratio is equal to 0.98 while

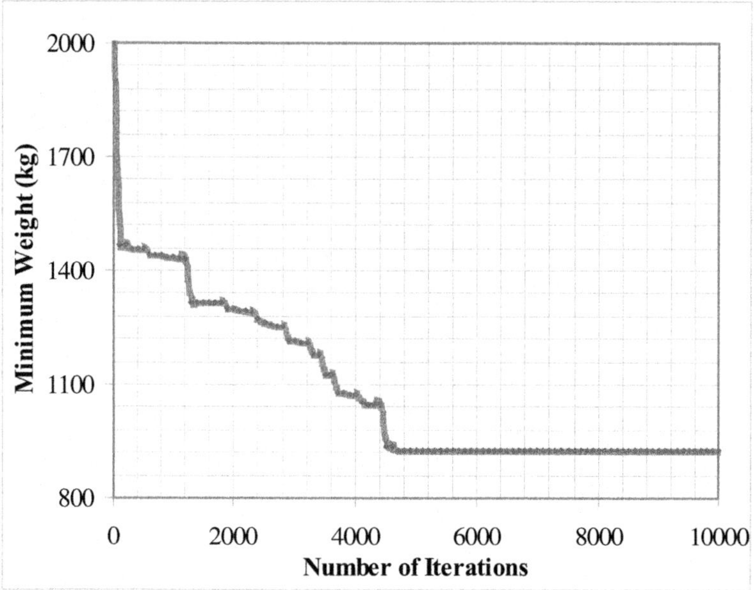

Fig. 13. The design history the cellular beam with 9m span

the maximum value of deflection of the beam is much smaller than its upper bound. This indicates that the strength constraints are dominant in the design problem. The optimum shape of the cellular beam obtained by the design algorithm under the loading considered is shown in Figure 12. The design history of this run is shown in Figure 13.

7 Conclusions

In this chapter it is shown that the optimum design problems of steel skeleton structures when formulated according to steel design codes of practice turn out to be combinatorial optimization problems where design variables are required to be selected from a discrete set. This feature of the programming problems is not affected by the code of practice adopted in the design. In the formulation of the optimum design of moment resisting steel frames, the optimum geometry design of geodesic domes and optimum design of cellular beams, British Code of Practice is considered while in the mathematical modeling of the optimum design problem of grillage systems the American steel design code is adopted. The solution of all these combinatorial structural design problems is obtained using the harmony search algorithm. In each of the design problems considered harmony search method attained the optimum designs without any adversity. The algorithm does not require gradient computations of the objective function and the constraints and its computer implementation is quite easy. It is observed that harmony search method is an effective and robust technique in finding the solution of combinatorial structural optimization problems. It can be stated with confidence that harmony search method provides a powerful alternative in developing optimum design algorithms for steel skeleton structures.

Acknowledgement

Author would like to thank Dr. Z. W. Geem for providing the computer code for the harmony search method.

References

1. Atrek, E.: New directions in optimum structural design. John Wiley & Sons, New York (1984)
2. Dréo, J., Petrowski, A., Siary, P., Taillard, E.: Metaheuristics for hard optimization; methods and case studies. Springer, Berlin (2006)
3. Nocedal, J., Wright, S.J.: Numerical optimization. Springer, Berlin (1999)
4. Chong, E.K.P., Żak, S.H.: An introduction to optimization. John Wiley and Sons, New York (2001)
5. Sun, W., Yuan, Y.-X.: Optimization theory and methods; nonlinear programming. Springer, Berlin (2006)
6. Arora, J.S.: Methods for discrete variable structural optimization. In: Burns, S.A. (ed.) Recent Advances in Optimum Structural Design. ASCE, Reston (2002)
7. Horst, R., Tuy, H.: Global optimization; deterministic approaches. Springer, Berlin (1995)
8. Paton, R.: Computing with biological metaphors. Chapman & Hall, London (1994)
9. Adami, C.: An Introduction to artificial life. Springer, Berlin (1998)
10. Matheck, C.: Design in nature: learning from trees. Springer, Berlin (1998)
11. Mitchell, M.: An introduction to genetic algorithms. MIT Press, Cambridge (1998)
12. Kennedy, J., Eberhart, R., Shi, Y.: Swarm intelligence. Morgan Kaufmann Publishers, San Fransisco (2001)
13. Kochenberger, G.A., Glover, F.: Handbook of metaheuristics. Kluwer Academic Publishers, USA (2003)
14. De Castro, L.N., Von Zuben, F.J.: Recent developments in biologically inspired computing. Idea Group Publishing, USA (2005)
15. Saka, M.P.: Optimum design of steel frames using stochastic search techniques based on natural phenomena: a review. In: Topping, B.H.V. (ed.) Civil Engineering Computations: Tools and Techniques. Saxe-Coburgh Publications, Scotland (2007)
16. Geem, Z.W., Kim, J.H., Loganathan, G.V.: A new heuristic optimization algorithm: harmony search. Simulation 76, 60–68 (2001)
17. Geem, Z.W., Kim, J.H., Loganathan, G.V.: Harmony search optimization: application to pipe network design. International Journal of Modeling and Simulation 22, 125–133 (2002)
18. Lee, K.S., Geem, Z.W.: A new structural optimization method based on harmony search algorithm. Computers and Structures 82, 781–798 (2004)
19. Lee, K.S., Geem, Z.W.: A new meta-heuristic algorithm for continuous engineering optimization: harmony search theory and practice. Computer Methods in Applied Mechanics and Engineering 194, 3902–3933 (2005)

20. Geem, Z.W., Lee, K.S., Tseng, C.-L.: Harmony search for structural design. In: Proceedings of 2005 Genetic and Evolutionary Computation Conference (GECCO 2005), Washington, USA, pp. 651–652 (2005)
21. Geem, Z.W.: Optimal cost design of water distribution networks using harmony search. Engineering Optimization 38, 259–280 (2006)
22. British Standards, BS 5950. Structural use of steelworks in building. part 1. code of practice for design in simple and continuous construction, hot rolled sections. British Standard Institution, London (1990)
23. McGuire, W.: Steel structures. Prentice Hall, Englewood Cliffs (1968)
24. Saka, M.P.: Optimum design of steel sway frames to BS 5950 using harmony search algorithm. Journal of Constructional Steel Research (in press, 2008)
25. Steelwork design guide to BS 5950: part 1, Section properties, member capacities, 4th edn., vol. 1. The Steel Construction Institute, London (1990)
26. Saka, M.P.: Optimum design of skeletal structures: a review. In: Topping, B.H.V. (ed.) Progress in Civil and Structural Engineering Computing, pp. 237–284. Saxe-Coburg Publications, Scotland (2003)
27. LRFD-AISC manual of steel construction. Load and resistance factor design (Second edn.). American Institute of Steel Construction (1999)
28. Saka, M.P., Erdal, F.: Harmony search based optimum design grillages. Strucutural and Multidisciplinary Optimization (in press, 2008)
29. Erdal, F., Saka, M.P.: Effect of beam spacing in the harmony search based optimum design of grillages. Asian Journal of Civil Engineering 9, 215–228 (2008)
30. Makowski, Z.S.: Analysis, design and construction of braced domes. Granada Publishing Ltd., London (1984)
31. Saka, M.P.: Optimum geometry design of geodesic domes using harmony search algorithm. Advances in Structural Engineering 10, 595–606 (2007)
32. Macsteel Service Centres (2008) (accessed December 8, 2008), `http://www.macsteel.zo.ca/cellbeam`
33. Ward, J.K.: Design of composite and non-composite cellular beams. The Steel Construction Institute Publication, London (1990)
34. Erdal, F., Saka, M.P.: Optimum design of cellular beams using the harmony search method. In: Proceedings of the Ninth International Conference on Computational Structures Technology. Civil-Comp Press, Stirlings (2008)

Harmony Search Algorithms for Water and Environmental Systems

Zong Woo Geem[1], Chung-Li Tseng[2], and Justin C. Williams[3]

[1] Environmental Planning and Management Program, Johns Hopkins University, Baltimore, Maryland, USA
zwgeem@gmail.com
[2] Australian School of Business, University of New South Wales, Sydney, Australia
c.tseng@unsw.edu.au
[3] Department of Geography and Environmental Engineering, Johns Hopkins University, Baltimore, Maryland, USA
jcwjr@jhu.edu

Abstract. Recently, the harmony search (HS) algorithm and other phenomenon-inspired algorithms have gained attention for their abilities to solve large-scale, difficult combinatorial optimization problems. This chapter reviews the applications of the HS method in the areas of water resources and environmental system optimization. Four specific optimization problems are considered: design of water distribution networks, scheduling of multi-location dams, parameter calibration of environmental models, and determination of ecological reserve location. The computational performance of the HS method on solving these four optimization problems will be compared against other optimization methods. It will be shown in this chapter that the HS method can outperform other methods in terms of solution quality and computational time.

Keywords: Harmony Search, Water Network Design, Dam Operation, Model Parameter Calibration, Ecological Optimization.

1 Introduction

This chapter presents four important environmental applications that can be tackled using the Harmony Search (HS) algorithm. These four applications include design of clean water networks, operation of multi-purpose dams, parameter calibration of a flood model, and determination of ecological reserve locations.

With the rise of environmental consciousness, there is growing concern on effectiveness and environmental impact of public infrastructure. In 2007, the British Medical Association, one of the most prestigious medical societies, announced that sanitation including sewage disposal and clean water distribution was the greatest medical milestone during the last century, beating other candidates including the inventions of antibiotics, vaccines and anesthesia [1]. This highlighted the indispensable nature of water distribution networks in our daily lives. Without a sound water distribution network to maintain sanitation, public health would be at risk. How to design an efficient and cost-effective water network using the HS algorithm is the first application covered in this chapter.

A dam is a barrier structure with various purposes, such as flood control, drought management, irrigation, water consumption, and electric power generation. The second

Z.W. Geem (Ed.): Music-Inspired Harmony Search Algorithm, SCI 191, pp. 113–127.
springerlink.com

example of the HS algorithm introduced in this chapter is about dam operations that maximize the benefits from irrigation and electricity generation while satisfying the operation constraints on release and storage amounts.

The HS algorithm was also applied to the parameter calibration of a flood routing model. Three parameters of the model are calibrated so that the model can accurately predict real-world flood amounts along the channel during certain time frame. This is the third HS application covered in this chapter.

The last HS application in this chapter is biological conservation. As more lands are developed, one of the costs is a loss of wildlife habitat [2]. It will be shown how biological diversity can be achieved efficiently using HS. Particularly, the maximal covering species problem (MCSP) is modeled to select a set of reserve sites that maximizes the number of species protected for a given budget or level of effort.

2 Design of Water Distribution Networks

A water distribution network (WDN) is one critical infrastructure that conveys water from sources to end consumers without any infection. The network consists of various interconnected hydraulic components, such as pipes, pumps, tanks, reservoirs, treatment plants, etc.

Three key factors, policies, conservation strategies, and design, must be considered in developing a WDN. Policies establish the service objectives and financing methods; conservation strategies focus on conserving schemes and monitoring systems; and design determines system layout and facility size [3].

Policies can be applied to service area which is defined as the geographical boundaries, with consideration of two major factors (level of service and financing). The level of service refers to the adequacy and reliability of service provided to customers, and it includes the factors such as adequate pressure, adequate amount (maximum-hourly demand), adequate hydrant amount, and reliability. Financing mechanisms include raw-water development fees, treated-water system fees, rates, bond issues, and taxes.

Water conservation has recently gained increasing awareness because of water shortage. Traditionally, engineers designed water network systems to meet required demand without considering the efficiency of the systems. In recent years, however, conservation groups have promoted more efficient water systems, as an effort to conserve water. In light of conservation efforts, pipe condition assessment is important because it can make engineers easier to find leak or break points of pipelines [4].

Several factors are considered for water network design, such as forecasted water demand, hydraulic criteria, network layout, and network size. In terms of future water demand, it can be estimated based on future population, which can be forecasted using various models such as arithmetical progression, geometrical progression, and exponential method [5]. Hydraulic criteria are engineering design values that ensure an adequate level of service, including maximum and minimum pressure, pressure fluctuation, head loss in pipes, and maximum velocity. Network layout and the size (pipe diameter or pump power) of a tree-like WDN whose capacity decreases along the distance from the source node can be determined using optimization techniques.

In this section, we focus on determining the size of a WDN such that it is cost-efficient. Our approach is compared against some benchmarks of real examples.

2.1 Problem Formulation for Water Network Sizing Design

The objective function of the water network design is generally the total costs of pipes and pumps that are functions of diameter sizes and pumping capacities. When certain design constraints are violated, such as minimal nodal pressure [6], penalty is imposed to the objective function to improve feasibility.

There are three major design constraints considered: mass conservation, energy conservation, and minimal nodal pressure. For every node, total inflows equal to total outflows in order to conserve the mass. For every loop, the summation of energy loss due to friction between fluid and metal is assumed to be zero (note an energy loss may take a positive or negative sign depending on its flow direction). For every demand node, the residual pressure should be greater than or equal to the minimal nodal pressure. The energy loss is calculated using EPANET [7].

2.2 Two-Loop Benchmark Network Example

Alperovits and Shamir [8] first proposed the benchmark example of two-loop network which has one reservoir, six demand nodes, and eight pipes, as shown in Figure 1.

The number of total candidate designs is $14^8 = 1.48 \times 10^9$ because the number of pipes (= decision variables) is 8 and the number of candidate diameters for each pipe is 14.

Table 1 shows the computational results of various meta-heuristic algorithms such as genetic algorithm (GA) [9], simulated annealing (SA) [10], shuffled frog leaping (SFL) [11], cross entropy (CE) [12], scatter search (SS) [13], and HS [14].

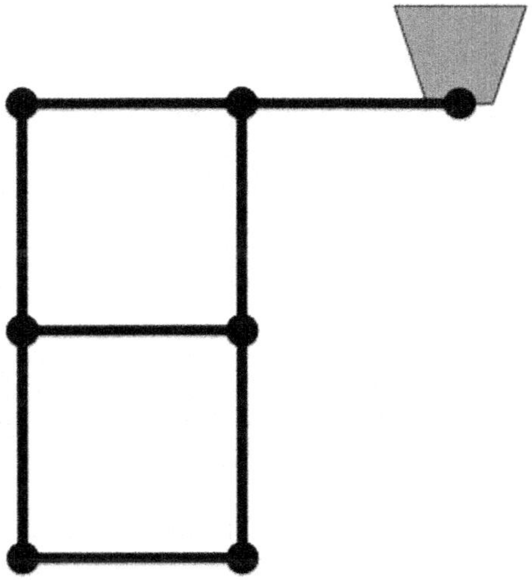

Fig. 1. Schematic of two-loop benchmark network

Results show that HS reached the optimal design solution (cost = $419,000) with the least number of function evaluations among competing algorithms, taking less than one second to solve using a desktop computer with an Intel Celeron 1.8GHz CPU. It should be also noted that GA and SA used less-strict hydraulic coefficients, which makes it much easier for both algorithms to find the optimal solution.

Table 1. Design Results of Two-Loop Network

Algorithms	Minimal Cost	Number of Evaluations
Genetic Algorithm	$ 419,000	7,467
Simulated Annealing	$ 419,000	> 25,000
Shuffled Frog Leaping	$ 419,000	11,155
Cross Entropy	$ 419,000	35,000
Scatter Search	$ 419,000	3,215
Harmony Search	$ 419,000	1,121

2.3 Hanoi Network Example

Fujiwara and Khang [15] first proposed the real-world example of Hanoi network located in Vietnam with a reservoir, 31 demand nodes, and 34 pipes, as shown in Figure 2.

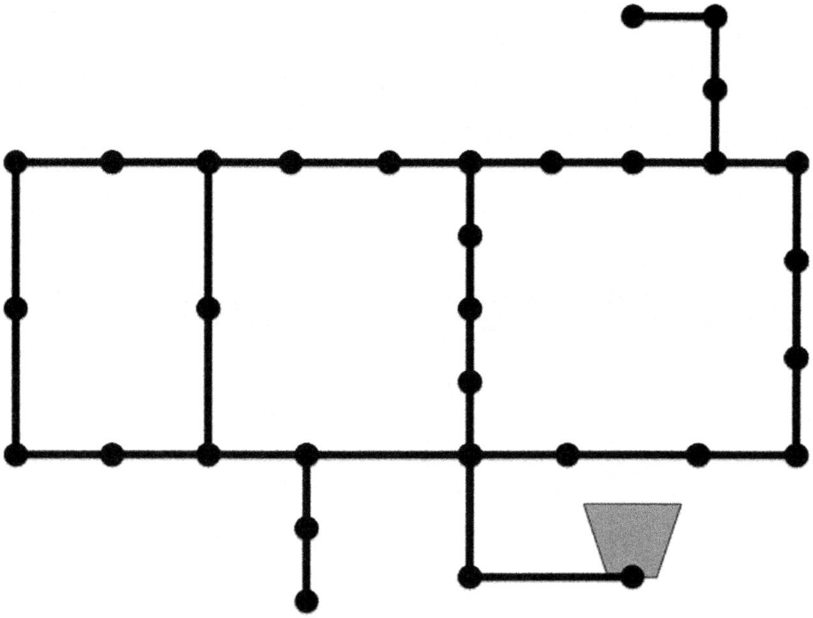

Fig. 2. Schematic of Hanoi network

The number of total candidate designs is $6^{34} = 2.87 \times 10^{26}$, which is derived because the number of pipes is 34 and the number of candidate diameters for each pipe is 6.

Table 2 shows the computational results using various meta-heuristic algorithms such as GA [16], ant colony optimization (ACO) [17], CE [12], SS [13], and HS [14]. Results show that HS reached the optimal design solution (cost = $ 6.018 million) with the least number of function evaluations comparing with other algorithms. It takes 20 seconds to solve this problem using the same desktop computer mentioned

Table 2. Design Results of Hanoi Network

Algorithms	Minimal Cost	Number of Evaluations
Genetic Algorithm	$ 6.173 M	26,457
Ant Colony Optimization	$ 6,134 M	35,433
Cross Entropy	$ 6.081 M	97,000
Scatter Search	$ 6.081 M	43,149
Harmony Search	$ 6.081 M	27,721

previously with an Intel Celeron 1.8GHz CPU. It should be noted that GA and ACO did not achieve the best objective function value found. The tabu search (TS) [18] result also violated the design constraint, which has been verified by EPANET [7].

2.4 Balerma Network Example

Reca and Martinez [19] proposed the real-world example of Balerma network located in Spain with 4 reservoirs, 443 demand nodes, and 454 pipes, as shown in Figure 3.

Fig. 3. Schematic of Balerma network

The total number of candidate designs is 10^{454}, where 454 is the number of pipes and 10 is the number of candidate diameters for each pipe.

Table 3. Design Results of Balerma Network

Algorithms	Minimal Cost	Number of Evaluations
Genetic Algorithm	€ 3.738 M	45,400
Simulated Annealing	€ 3.476 M	45,400
SA + TS	€ 3.298 M	45,400
Harmony Search	€ 2.601 M	45,400

Table 3 shows the computational results of various meta-heuristic algorithms such as GA [16], SA [16], simulated annealing & tabu search (SA+TS) [16], and HS [14]. Results show that HS reached the least-cost design solution (cost = € 2.601 million) with the same number of function evaluations, taking eight minutes on a desktop computer with Intel Celeron 1.8GHz CPU.

2.5 New York City Network Example

Schaake and Lai [20] proposed the real-world example of New York City network which has one reservoir, 19 demand nodes, and 21 pipes, as shown in Figure 4.

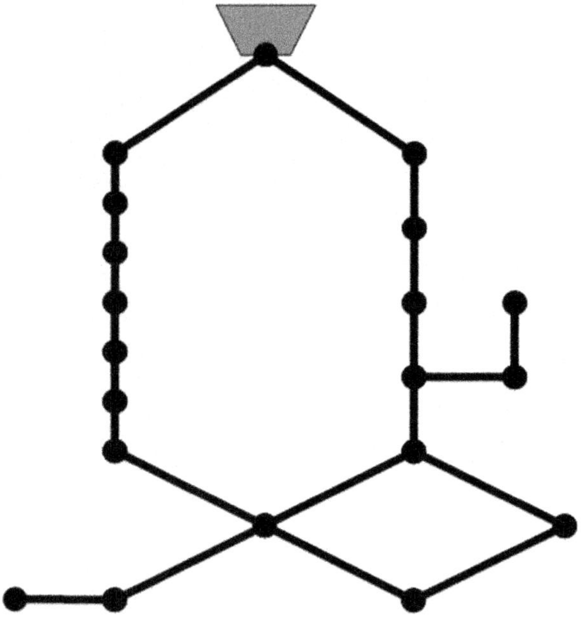

Fig. 4. Schematic of New York City network

The total number of candidate designs is $16^{21} = 1.93 \times 10^{25}$, where 21 is the number of parallel pipes and 16 is the number of candidate diameters for each parallel pipe.

Table 4 shows the computational results of various meta-heuristic algorithms such as GA [21], ACO [22], SFL [11], and HS [14]. Results show that HS reached the least-cost design solution (cost = \$ 38.64 million) with the least number of function evaluations, taking three seconds on a desktop computer with Intel Celeron 1.8GHz CPU.

Table 4. Design Results of New York City Network

Algorithms	Minimal Cost	Number of Evaluations
Genetic Algorithm	\$ 38.64 M	800,000
Ant Colony Optimization	\$ 38.64 M	7,014
Shuffled Frog Leaping	\$ 38.80 M	21,569
Harmony Search	\$ 38.64 M	3,373

2.6 Other Issues

In the previous sections, how the HS algorithm was applied to the design of water distribution networks has been shown. Recently another study showed how HS can be applied to the design of a pump-included water network [23]. In the study, HS found the identical least-cost solution with less number of function evaluations when compared with SA.

HS was also applied to the layout design of tree-type water networks [24, 25]. For the layout problem of 64-node benchmark network reported in [24] that consists of 1.26×10^{25} possible combinations in total, HS reached the global optimum after only 1,500 evaluations while GA [26] and evolutionary algorithm (EA) [27] reached near-optima after 3,200 evaluations. For the layout problem of 100-node real-world network in [25] that consists of 3.65×10^{54} possible combinations, HS reached a near-optimum within 0.4% of the optimal objective value, while EA [27] reached another near-optimum within 1.0% and ACO [28] within 11.4%.

Recently, HS has been further improved by incorporating particle swarm technique [14]. Results showed that particle-swarm HS converges much faster than the original HS especially for small or medium sized networks such as two-loop network and Hanoi network.

3 Scheduling of Multiple Dams

A dam is a barrier structure retaining water for the purpose of irrigation, urban water supply, navigation, industrial uses, hydroelectric power generation, recreation uses, wildlife habitat creation, and flood control [29].

For the optimal scheduling of multiple-unit, multiple-purpose dam systems, traditionally researchers have used dynamic programming techniques [30] to solve it. The dynamic programming approach suffers from the so-called curse of dimensionality, which results in large memory requirement and long computing time. Thus, researchers have diverted their attention to meta-heuristic algorithms such as GA [31], SA [32], and HS [33].

To test the algorithm performance, a four-dam system in Figure 5, first proposed by Larson [34], is considered.

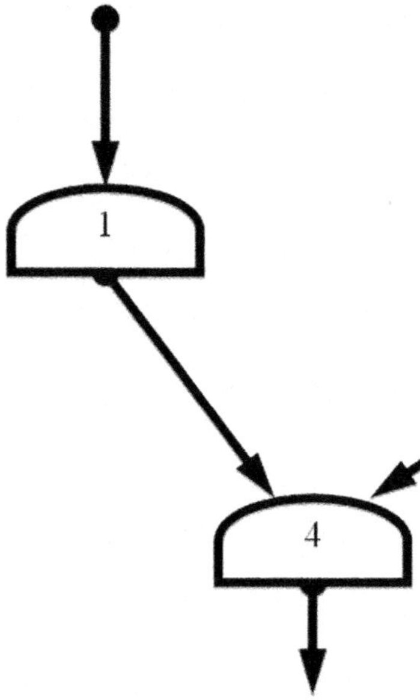

Fig. 5. Schematic of four-dam system

The objective of the multiple dam scheduling optimization is to maximize the benefits from hydroelectric power generation and irrigation while satisfying all operational constraints such as range of water releases, range of dam storages, and mass conservation between inflows and outflows. Boundary conditions such as initial storages and final storages must also be satisfied.

Although the total number of candidate schedules is 6.87×10^{34}, HS found five different global optima with the identical maximal benefits after 35,000 evaluations, taking 46 seconds on a desktop with Intel Celeron 1.8GHz CPU [33]. However, GA (with binary, gray, & real-value representations; tournament selection; one-point, two-point, and uniform crossovers; and uniform and modified uniform mutations) only reached a near-optimum after the same amount of evaluations [31].

4 Parameter Calibration of Flood Routing Model

A flood is a water overflow that submerges land [35]. Flooding results from the situation when the volume of a water body, such as a river or lake, exceeds the total capacity of its bounds. Because the flood causes many problems, such as infrastructure damage, casualties, epidemics, contamination of water, shortage of harvest, and economic hardship, it is important to predict the flood amount along the timespan.

One popular hydrologic model that routes the flood is Muskingum model (Figure 6) originally developed by McCarthy [36] after his research of the Muskingum river basin in Ohio. The Muskingum model had two hydrologic parameters. An additional parameter was added by Gill [37] in order to represent the characteristic of the flood more accurately. Up to now, various techniques, such as least-square method (LS) [37], modified Hooke-Jeeves (HJ) pattern search [38], nonlinear technique (NL) [39], GA [40], HS [41], Lagrange multiplier (LM) method [42], and Broyden-Fletcher-Goldfarb-Shanno (BFGS) method [43], have tackled the optimal parameter calibration of the three-parameter Muskingum model. The objective of the approach is to minimize the difference between real-world flood amounts and computed ones in terms of the residual sum of squares (SSQ).

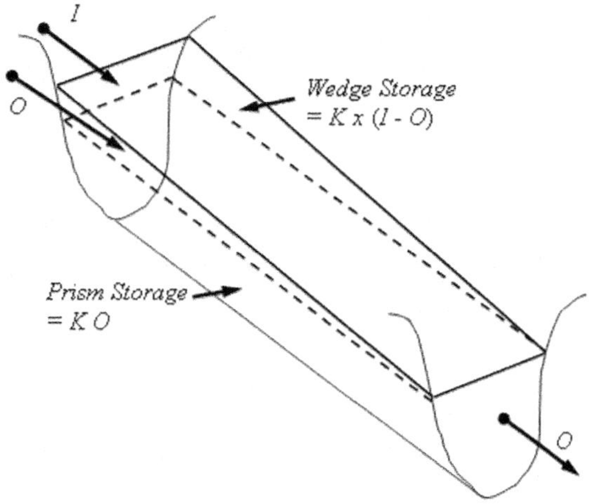

Fig. 6. Schematic of Muskingum flood routing model

Table 5. Parameter Calibration Results of Muskingum Model

Algorithms	SSQ
LS	145.69
HJ	45.61
NL	156.44
GA	38.24
HS[1]	36.78
HS[2]	36.77
LM	130.49
BFGS	36.77

Table 5 shows the computational results of various algorithms. HS found the near-optima (SSQ = 36.78 and 36.77) that are very close to global optimum reached by BFGS (SSQ = 36.77). In Table 5, HS^1 indicates the discrete-version of the HS algorithm [41], and HS^2 is a continuous version [44].

Although the BFGS was able to arrive at the global optimum, only 26.7% of the randomly generated starting parameter values reach this level. A hybrid model, recently reported in [45], that integrated HS and gradient-based algorithms may enhance the performance of the parameter calibration problem. This will be investigated in the future.

5 Planning of Biological Conservation

Excessive urban area expansion is increasingly causing the loss or degradation of wildlife habitat. The protection of wildlife in the face of urban growth and other development pressures is an important and challenging task in biological conservation. If current trends continue, precious species of flora and fauna will be subject to growing risks of extinction. The HS algorithm may be used to help protect wildlife in general, including at-risk species by identifying effective and efficient plans for saving critical habitat [46].

There are at least five different optimization models [47] for optimally selecting ecological reserves: species set covering problem (SSCP), maximal covering species problem (MCSP), maximal multiple-representation species problem (MMRSP), chance constrained covering problem, and expected covering problem.

Among these five models, MCSP, which maximizes the number of species conserved, while limiting the number of reserves operated, is one of the more commonly used models. The MCSP is formulated as follows:

$$\text{Maximize} \sum_{i \in I} y_i \tag{1}$$

$$s.t. \sum_{j \in M_i} x_j \ge y_i, \quad all\; i \in I, \tag{2}$$

$$\sum_{j \in J} x_j = P, \tag{3}$$

where i and I are the index and set of species, respectively; y_i is a binary variable representing whether species i is covered or not (it is one if species i is covered); j and J are the index and set of candidate reserves, respectively; M_i is the set of candidate reserves that include species i; x_j is a binary variable for reserve selection (it is one if candidate reserve j is selected); and P is the number of candidate reserves to be selected.

The MCSP model was applied to real-world ecological conservation problem in Oregon, USA, as shown in Figure 7. Total 426 vertebrate species were identified for this data set, which consists of 441 hex parcels (candidate reserves). Each parcel contains one or more of the 426 listed species [48].

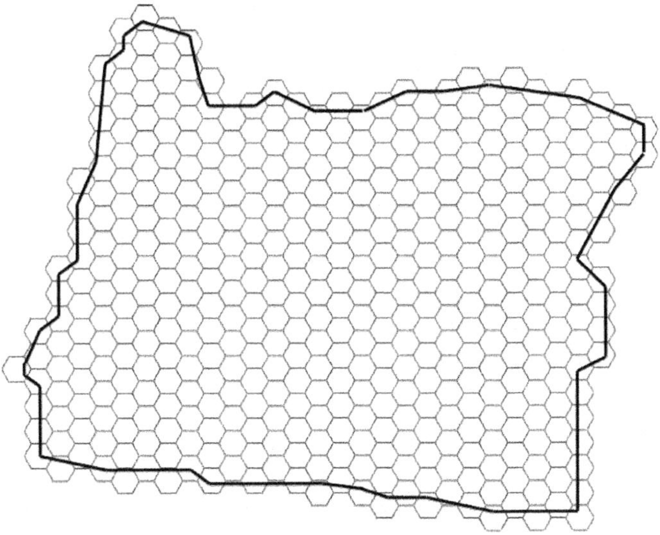

Fig. 7. Oregon Hex Map

Table 6. Comparison of Oregon MCSP Results

P	SA	HS	Opt	*P*	SA	HS	Opt
1	254	254	254	14	413	414	414
2	318	318	318	15	415	416	416
3	356	356	356	16	417	417	418
4	374	374	374	17	418	419	419
5	384	384	384	18	419	420	420
6	390	390	390	19	420	421	422
7	395	395	395	20	421	422	423
8	398	400	400	21	422	423	424
9	403	402	403	22	423	424	425
10	405	405	406	23	424	425	426
11	407	408	408	24	425	426	NA
12	409	410	410	25	426	NA	NA
13	411	412	412	-	-	-	-

In order to optimally solve the MCSP, the original HS algorithm was modified by adopting several problem-specific features such as sparse selection, big-reserve-first selection, and diversity-first selection. In sparse selection, each binary variable x_j for reserve acquisition is selected with much less probability than 50% because only P reserves should be selected out of 441 candidates; in big-reserve-first selection, a reserve that has more species, has earlier chance to be selected; and in diversity-first selection, a reserve that represents the only occurrence of a species, is selected unconditionally. Also, because each decision variable has binary value, there is no pitch adjustment.

Table 6 shows the computational results of SA [49], HS [46], and the exact method (Opt) [50]. As seen in the table, the modified HS reached global optimum in 15 cases out of 24 cases. When compared with SA, HS reached better solutions in 14 cases while it reached worse solution only once.

Although HS did not reach the global optimum (all 426 species covered) with P = 23, it was able to find many different near-optima (426) with P = 24. As a practical matter, the identification of alternate optima is valuable in planning and decision making; if a particular reserve became unavailable for some reason, another solution that utilized different parcels but had the same payoff could (potentially) be implemented.

6 Summary and Conclusions

This chapter has reviewed various HS applications in the areas of water resources and environmental planning and management. Specific topics include design of water distribution networks, scheduling of multiple dam system, parameter estimation of flood routing model, and determination of ecological reserve locations.

The computational results showed that HS has comparative advantages in clean water distribution, irrigation, hydroelectric power generation, flood prediction, and nature conservation. For clean water distribution, HS found the least-cost design with the least number of function evaluations when compared with other methods. In terms of maximal benefit from irrigation and power generation, HS found multiple global optima while others only reached near-optima. To better predict flood amount using hydrologic model, HS calibrated three hydrologic parameters and minimized the difference between observed and routed data, and outperformed other methods. Finally, for nature conservation HS was modified by adopting several problem-specific operators and in many cases obtained more effective conservation plans than another existing heuristic method (SA) in literature.

With the success of these applications, our future research direction focuses on further promoting the HS algorithms to more research areas. Particularly, we will look into how to integrate HS with various practical technologies such as Web, GUI, DB, GIS, and API.

References

1. BBC News, Sanitation 'best medical advance' (2007) (accessed October 20, 2008), `http://news.bbc.co.uk/2/hi/health/6275001.stm`
2. Rosing, K.E., ReVelle, C.S., Williams, J.C.: Maximizing species representation under limited resources: a new and efficient heuristic. Environmental Modeling and Assessment 7, 91–98 (2002)
3. Cesario, L.: Modeling, analysis, & design of water distribution systems. American Water Works Association, Denver (1995)
4. Geem, Z.W., Tseng, C.L., Kim, J., et al.: Trenchless water pipe condition assessment using artificial neural network. In: Najafi, M., Osborn, L. (eds.) Pipelines 2007: advances and experiences with trenchless pipeline projects. ASCE, Reston (2007)

5. Choi, W.S.: Water supply and sewerage. ChungMunGak, Seoul (1999)
6. Geem, Z.W.: Optimal cost design of water distribution networks using harmony search. Engineering Optimization 38, 259–280 (2006)
7. Rossman, L.A.: EPANET2 users manual. US Environmental Protection Agency, Cincinnati (2000)
8. Alperovits, E., Shamir, U.: Design of optimal water distribution systems. Water Resources Research 13, 885–900 (1977)
9. Wu, Z.Y., Boulos, P.F., Orr, C.H., et al.: Using genetic algorithms to rehabilitate distribution systems. Journal of the American Water Works Association 93, 74–85 (2001)
10. Cunha, M.C., Sousa, J.: Water distribution network design optimization: simulated annealing approach. ASCE Journal of Water Resources Planning and Management 125, 215–221 (1999)
11. Eusuff, M., Lansey, K.E.: Optimization of water distribution network design using the shuffled frog leaping algorithm. ASCE Journal of Water Resources Planning and Management 129, 210–225 (2003)
12. Perelman, L., Ostfeld, A.: An adaptive heuristic cross-entropy algorithm for optimal design of water distribution systems. Engineering Optimization 39, 413–428 (2007)
13. Lin, M.D., Liu, Y.H., Liu, G.F., et al.: Scatter search heuristic for least-cost design of water distribution networks. Engineering Optimization 39, 857–876 (2007)
14. Geem, Z.W.: Particle-swarm harmony search for water network design. Engineering Optimization (in press)
15. Fujiwara, O., Kang, D.B.: A two-phase decomposition method for optimal design of looped water distribution networks. Water Resources Research 26, 539–549 (1990)
16. Reca, J., Martinez, J., Gil, C., et al.: Application of several meta-heuristic techniques to the optimization of real looped water distribution networks. Journal Water Resources Management 22, 1367–1379 (2008)
17. Zecchin, A.C., Simpson, A.R., Maier, H.R., et al.: Application of two ant colony optimization algorithms to water distribution system optimization. Mathematical and Computer Modelling 44, 451–468 (2006)
18. Cunha, M.C., Ribeiro, L.: Tabu search algorithms for water network optimization. European Journal of Operational Research 157, 746–758 (2004)
19. Reca, J., Martinez, J.: Genetic algorithms for the design of looped irrigation water distribution networks. Water Resources Research 42 (2006)
20. Schaake, J., Lai, D.: Linear programming and dynamic programming – application of water distribution network design (Report 116). MIT Press, Cambridge (1969)
21. Broad, D.R., Dandy, G.C., Maier, H.R.: Water distribution system optimization using metamodels. ASCE Journal of Water Resources Planning and Management 131, 172–180 (2005)
22. Maier, H.R., Simpson, A.R., Zecchin, A.C., et al.: Ant colony optimization for design of water distribution systems. ASCE Journal of Water Resources Planning and Management 129, 200–209 (2003)
23. Geem, Z.W.: Harmony search optimisation to the pump-included water distribution network design. Civil Engineering and Environmental Systems (2008), doi:10.1080/10286600801919813
24. Geem, Z.W., Park, Y.: Harmony search for layout of rectilinear branched networks. WSEAS Transactions on Systems 6, 1349–1354 (2006)
25. Geem, Z.W.: Geometry layout for real-world tree networks using harmony search. In: Proceedings of the 3rd Indian International Conference on Artificial Intelligence (IICAI 2007), pp. 268–277 (2007)

26. Walters, G., Lohbeck, T.: Optimal layout of tree networks using genetic algorithms. Engineering Optimization 22, 27–48 (1993)
27. Walters, G., Smith, D.: Evolutionary design algorithm for optimal layout of tree networks. Engineering Optimization 24, 261–281 (1995)
28. Afshar, M.H., Marino, M.A.: Application of ant algorithm for layout optimization of tree networks. Engineering Optimization 38, 353–369 (2006)
29. Wikipedia, Dam (2002) (accessed October 23, 2008), `http://en.wikipedia.org/wiki/Dam`
30. Heidari, M., Chow, V.T., Kokotovic, P.V., et al.: Discrete differential dynamic programming approach to water resources systems optimization. Water Resources Research 7, 273–282 (1971)
31. Wardlaw, R., Sharif, M.: Evaluation of genetic algorithms for optimal reservoir system operation. ASCE Journal of Water Resources Planning and Management 125, 25–33 (1999)
32. Teegavarapu, R.S.V., Simonovic, S.P.: Optimal operation of reservoir systems using simulated annealing. Water Resources Management 16, 401–428 (2002)
33. Geem, Z.W.: Optimal scheduling of multiple dam system using harmony search algorithm. In: Sandoval, F., Prieto, A.G., Cabestany, J., Graña, M. (eds.) IWANN 2007. LNCS, vol. 4507, pp. 316–323. Springer, Heidelberg (2007)
34. Larson, R.E.: State increment dynamic programming. American Elsevier, New York (1968)
35. Wikipedia, Flood (2002) (accessed October 23,2008), `http://en.wikipedia.org/wiki/Flood`
36. McCarthy, G.T.: The unit hydrograph and flood routing. In: Proceedings of Conference of the North Atlantic Division. U.S. Army Corps of Engineers, New London (1938)
37. Gill, M.A.: Flood routing by the Muskingum method. Journal of Hydrology 36, 353–363 (1978)
38. Tung, Y.K.: River flood routing by nonlinear Muskingum method. ASCE Journal of Hydraulic Engineering 111, 1447–1460 (1985)
39. Yoon, J., Padmanabhan, G.: Parameter estimation of linear and nonlinear Muskingum models. ASCE Journal of Water Resources Planning and Management 119, 600–610 (1993)
40. Mohan, S.: Parameter estimation of nonlinear Muskingum models using genetic algorithm. ASCE Journal of Hydraulic Engineering 123, 137–142 (1997)
41. Kim, J.H., Geem, Z.W., Kim, E.S.: Parameter estimation of the nonlinear Muskingum model using harmony search. Journal of the American Water Resources Association 37, 1131–1138 (2001)
42. Das, A.: Parameter estimation for Muskingum models. ASCE Journal of Irrigation and Drainage Engineering 130, 140–147 (2004)
43. Geem, Z.W.: Parameter estimation for the nonlinear Muskingum model using BFGS technique. ASCE Journal of Irrigation and Drainage Engineering 132, 474–478 (2006)
44. Lee, K.S., Geem, Z.W.: A new meta-heuristic algorithm for continuous engineering optimization: harmony search theory and practice. Computer Methods in Applied Mechanics and Engineering 194, 3902–3933 (2005)
45. Fesanghary, M., Mahdavi, M., Minary-Jolandan, M., et al.: Hybridizing harmony search algorithm with sequential quadratic programming for engineering optimization problems. Computer Methods in Applied Mechanics and Engineering 197, 3080–3091 (2008)
46. Geem, Z.W., Williams, J.C.: Harmony search and ecological optimization. International Journal of Energy and Environment 1, 150–154 (2007)

47. ReVelle, C., Williams, J.C., Boland, J.J.: Counterpart models in facility location science and reserve selection science. Environmental Modelling and Assessment 7, 71–80 (2002)
48. Masters, L., Clupper, N., Gaines, E., et al.: Biodiversity Research Consortium Species Database Manual, The Nature Conservancy, Boston (1995)
49. Csuti, B.S., Polasky, P.H., Williams, R.L., et al.: A comparison of reserve selection algorithms using data on terrestrial vertebrates in Oregon. Biological Conservation 80, 83–97 (1997)
50. Rosing, K.E., ReVelle, C.S., Williams, J.C.: Maximizing species representation under limited resources: a new and efficient heuristic. Environmental Modeling and Assessment 7, 91–98 (2002)

Identification of Groundwater Parameter Structure Using Harmony Search Algorithm

M. Tamer Ayvaz

Department of Civil Engineering, Pamukkale University, Denizli, Turkey
tayvaz@pamukkale.edu.tr

Abstract. Mathematical simulation models are widely used to predict the future response of groundwater systems for different flow and mass transport conditions. These models are based on the solution of governing partial differential equations which require some spatially distributed hydro-geological model parameters. However, these parameters are usually unknown due to the complexity of groundwater systems. Therefore, identification of them is an important task since they are the primary input of management models used in groundwater modeling. This chapter provides a brief review dealing with the solution of parameter structure identification problems based on the harmony search optimization algorithm. The results of this review indicate that the harmony search algorithm yields nearly same or better solutions than those of a genetic algorithm, which is another popular meta-heuristic optimization algorithm.

Keywords: Groundwater Management, Parameter Structure Identification, Simulation-Optimization, Harmony Search Algorithm.

1 Introduction

Groundwater is an important water resource throughout the world. Nowadays, the significance of the groundwater resources continuously grows as society and economics develop. However, uncontrolled and unplanned use of groundwater resources may cause some serious geological problems, including groundwater depletion, seawater intrusion, land subsidence and desertification [1]. Therefore, sustainable planning and management strategies should be developed to optimally operate the groundwater resources.

Mathematical simulation models are the essential tools of developing sustainable management strategies for groundwater systems. They are used to evaluate the aquifer's response for different management strategies [2] by solving governing flow and mass transport equations which require the spatial distributions of some hydrogeological parameters (hydraulic conductivity, transmissivity, storativity, etc.). Note that these parameters are usually determined by some laboratory or field studies, and determination of them for large groundwater systems may require significant time and cost. Therefore, much research has been conducted to determinate these parameter distributions based on field observations of hydraulic head values. This approach is identified as inverse modeling and usually solved using combined simulation/optimization (S/O) models [3].

It should be noted that combined S/O models have been employed to the solution of many groundwater modeling problems. These problems may be classified as three main groups: Groundwater parameter estimation problems; Groundwater hydraulic

Z.W. Geem (Ed.): Music-Inspired Harmony Search Algorithm, SCI 191, pp. 129–140.
springerlink.com

management problems; and Groundwater quality management problems. The first group deals with the solution of parameter estimation problems defined above. It aims to determine some hydro-geological parameter values through inverse modeling approach. The second group deals with the solution of groundwater hydraulic management problems. The typical problems for this group are to maximize the total yield of the aquifer system; or to minimize the total pumping costs to satisfy the given water demand [4]. In the last group, the main objective is to solve the groundwater quality management problems based on the historical observation data. Two typical problems for this group are to solve the contaminant source identification problems [5], and to determine the best remediation strategy to clean up the aquifer system [6]. Note that the groundwater parameter estimation problem described in the first group must be solved before solving the groundwater hydraulic and quality management problems since hydro-geological parameter distributions are required to solve the governing flow and transport equations.

The optimization problems dealing with the groundwater modeling usually have non-linear and non-convex solution spaces [7]. For such problems, finding a global optimum solution is not an easy task using traditional optimization algorithms since they require some derivatives and initial values [6-8]. Therefore, meta-heuristic algorithms are usually preferred for the solution of optimization problems dealing with groundwater modeling.

Recently, Geem et al. [9] developed the meta-heuristic Harmony Search (HS) optimization algorithm which is based on the musical process of searching for a perfect state of harmony. In the HS algorithm, musicians try to find the musically pleasing harmony by performing several improvisations. The quality of the generated harmonies is determined by aesthetic or artistic standards. In the optimization process, a global optimum solution may be found by performing several iterations through different values of decision variables. The quality of the obtained solutions is evaluated by an objective function. These are the main similarities between the musical improvisation and optimization. The main advantages of HS are: 1) HS does not require complex mathematical calculations; 2) Programming HS is simple; 3) HS does not require specifying the initial value settings; 4) HS can handle both discrete and continuous decision variables without requiring gradients.

The objective of this chapter is to review the application of the HS optimization algorithm to the solution of groundwater parameter structure identification problems. In order to evaluate the performance of the HS algorithm, two published papers were examined. The first paper [10] solves the parameter structure identification problem using a combinatorial optimization scheme in which a Genetic Algorithm (GA) is combined with a grid search and a quasi-Newton optimization algorithm. The second paper [11] solves the same problem using HS based optimization model. The obtained results indicate that HS requires fewer simulation runs and yields the same or better solutions than GA.

The remainder of this chapter is organized as follows: first, the main structure and computational structure of the HS algorithm is presented; second, how to apply the HS algorithm to the solution of inverse parameter structure identification problem is described; and finally, a comparison of HS and GA is performed.

2 The Meta-heuristic Harmony Search Algorithm

Like other meta-heuristic optimization algorithms, the main philosophy of the HS algorithm depends on a heuristic event. However, HS does not get its philosophy from the natural processes, instead, it gets it from the musical improvisation which occurs when a group of musicians searches for a better state of harmony. Geem et al. [9] first adapted this musical improvisation process into the solution of engineering optimization problems. In this adaptation, each musician corresponds to a decision variable and possible notes in the musical instruments correspond to the possible values of the decision variables. When the musicians find the fantastic harmony from their memories, it means, a global optimum solution is found through corresponding decision variables. Figure 1 demonstrates the analogous relationship the musical improvisation and optimization [12].

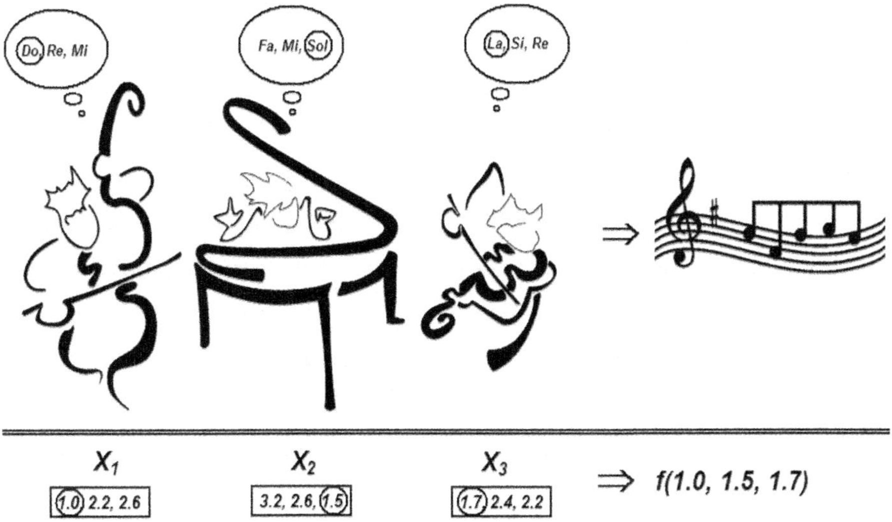

Fig. 1. Analogy between musical improvisation and optimization

As can be seen from Figure 1, each musician has several notes in their memories. For example, if the first musician plays {Do} while second and third musicians play {Sol} and {La} from their harmony memories, {Do, Sol, La} makes a new harmony. If this harmony is better than the worst one in the harmony memory, it replaces the worst one and this process is repeated until fantastic harmony is found. On the other hand, in the optimization process, each musician is replaced with a decision variable and notes are replaced with preferred values of decision variables. If the decision variables take {1.0}, {1.5}, and {1.7} from the harmony memory, a new solution vector {1.0, 1.5, 1.7} is obtained. If this solution vector is better than the worst one stored in the harmony memory, this new solution vector replaces the worst one. This iterative process is repeated until the given termination criterion is satisfied.

The computational scheme of the HS algorithm comes from three musical operations: i) playing a note from the harmony memory; ii) playing a note randomly from the entire note range; and iii) playing a note which is close to another one stored in memory. Combination of these operations allows finding a musically pleasing harmony. These operations can be adapted into the engineering optimization problems as follows: i) new generated variable values are selected from the harmony memory; ii) new variable values are randomly selected from the possible random range; iii) new variable values are further replaced with other values which are close to the values. Combination of these operations allows searching a global optimum solution in an optimization framework. For continuous decision variables, the computational scheme of HS can be given as follows [13]:

1. Generate vectors ($\mathbf{x}^1 \cdots \mathbf{x}^{HMS}$) randomly and put them into the memory.
2. Generate a new solution vector $\mathbf{x}'$ for each x_i':
   - With probability HMCR, select x_i' from memory, $x_i' = x_i^{Rnd(1,HMS)}$
   3. Adjust the pitch x_i' which is obtained from memory:
      - With probability PAR, change x_i' as, $x_i' = x_i' \pm bw \times Rnd(0;1)$.
      - With probability (1–PAR), do nothing.
   - With probability (1–HMCR), select x_i' randomly.
4. If $\mathbf{x}'$ is better than the worst $\mathbf{x}^j$ in harmony memory, replace $\mathbf{x}^j$ with $\mathbf{x}'$.
5. Repeat Steps 2 to 5 until the given termination criterion is satisfied.

The solution parameters of HS are: Harmony Memory Size (HMS), Harmony Memory Considering Rate (HMCR), Pitch Adjusting Rate (PAR), and distance bandwidth (*bw*).

The HS algorithm has recently been applied to various engineering optimization problems including music composition [14], Sudoku puzzle [15], structural design [16], water distribution network design [17], vehicle routing [18], dam scheduling [19], groundwater modeling [11, 20], soil stability analysis [21], energy system dispatch [22], and transport energy demand modeling [23], etc. The detailed information about the HS algorithm can be found at [24].

3 Parameter Structure Identification in Groundwater Modeling

This section defines the necessary solution steps for solving the parameter structure identification problem in groundwater modeling. This is an important problem since all the groundwater simulation models require these parameter structures while numerically solving the governing flow and mass transport equations. Aquifers are heterogeneous geological formations and distribution of their parameters is usually unknown. Hence, inverse modeling approaches are usually used to identify the associated parameter structures. Note that, inverse groundwater modeling problems often give unstable results due to ill-posedness which is characterized by the instability and non-uniqueness of the identified parameters. The instability and non-uniqueness is associated with the numerical or observational errors, possibility of the multiple solutions, as well as the limited field data [3, 11].

In groundwater modeling, usually two type parameterization approaches are used to identify the parameter structures: interpolation and zonation [25]. In the interpolation approach, the parameter structure of the flow domain is generated using some interpolation functions. However, this approach requires known parameter values at some locations to determine the whole parameter structure [26]. If there is no known parameter values available in the flow domain, this approach may not be used. On the other hand, in the zonation approach, the flow domain is subdivided into the zones where associated parameter values are assumed to be constant. This may be the more realistic approach if distribution of the aquifer parameters in the field is suitable for using the zonation approach.

Note that, identification of the zone structure of an aquifer system is more difficult than the identifying the associated parameter values within zones [27]. Hence, the number of the published studies dealing with the use of this approach is limited [3, 10-11, 20, 28-32].

In zonation approach, each zone includes a basis point location which is the key parameter of determining the shape of zone structure. Several zonation approaches have been proposed to characterize the parameter structures. One of the most widely used approaches is the Voronoi Tesselation (VT) [33]. In VT, zonation process is performed based on the metric distances between data points and basis points. VT has very simple computational procedure and always produces convex shaped zone structures. Another zonation approach is the Fuzzy c-Means Clustering (FCM) approach [34] in which each zone is considered as a fuzzy set and data points are assigned to the zones by a fuzzy membership grade. The key difference between VT and FCM is that VT uses the metric distance for assigning a data point into a zone whereas FCM uses fuzzy memberships [11]. Another version of the FCM algorithm is the kernel based FCM (KFCM). The main idea of the KFCM algorithm is to transform the original low-dimensional input space into a higher dimensional feature space. Through this transformation, more general zone shapes may be generated [3]. After partitioning the flow domain through zonation approach, homogeneous parameter values are assigned into each zone.

Note that inverse parameter structure identification problem can be solved through the combined S/O models. In these models, the locations of the basis points and the associated parameter values are treated as the decision variables of the optimization model. The objective of the optimization model is to identify the associated zone structure by minimizing the Residual Error (RE) between observed and simulated hydraulic heads at available observation wells. RE is used to identify the aquifer's response for each parameter zone structure as follows [11]:

$$\mathrm{RE}_C = Min\left[\sum_{t=1}^{N_T}\sum_{\mu=1}^{N_d}\left(h_\mu(\Omega_c,t)-\tilde{h}_\mu(t)\right)^2\right] \tag{1}$$

where RE_c is the residual error value in the parameter zone structure Ω_c, N_T is the total simulation time, N_d is the number of observation wells, h_μ and $\tilde{h}_\mu$ are the simulated and observed hydraulic heads in observation well μ.

Another important issue for the solution of the inverse parameter structure identification problem is the determination of the optimum zone structures. Although the combined S/O models are the efficient tools of identifying the parameter zone structures, in practical cases, the number of zones is also an unknown parameter to be determined. In order to identify the optimum zone structure, the inverse solution procedure starts with a zone and increases the zone number until the best zone structure is identified [10].

In this solution procedure, an increase in the zone numbers results with the decrease in the calculated RE values. However, this situation may result with the increase in the uncertainty of the identified parameters. In such cases, parameter structure identification process is stopped since the reliability of the identified parameters usually decreases. Therefore, it is necessary to use an additional measure, Parameter Uncertainty (PU), to determine the optimum zone structure. Note that PU is calculated based on the covariance matrix of the estimated parameters as follows [35]:

$$\text{Cov} = G_c \left(\frac{1}{N_d \times T - c} \sum_{t=1}^{T} \sum_{\mu=1}^{N_d} \left(h_\mu (\Omega_c, t) - \tilde{h}_\mu (t) \right)^2 \left[J^{\mathrm{T}} J \right]^{-1} \right) G_c^{\mathrm{T}} \tag{2}$$

where the superscript T is the transpose operation, J is the Jacobian matrix of hydraulic heads with respect to transmissivities, G_c is the structure matrix whose elements represent the weight of the parameter value form the ith basis point to the jth data point. After obtaining the covariance matrix, the trace or the norm of Eq. 2 is used to compute the PU values.

3.1 Model Application

Tsai et al. [10] first proposed the following hypothetical example, which is a two-dimensional confined aquifer system, to solve the parameter zone structure identification problem. Figure 2, which follows, shows the geometry and boundary conditions of the aquifer model under consideration.

It can be seen form Figure 2 that the boundary conditions of the aquifer is 100 m specified head at east, and no-flow at the other boundaries. Aquifer has 5 pumping wells (PWs) and it is assumed that all the wells are continuously operated for 10 days. The associated pumping rates of the wells are: 4,000 m^3/day for PW_1 to PW_4, and 2,000 m^3/day for PW_5. The storage coefficient is 0.0002. There are seven observation points available where head observations are measured at each day. In order to simulate the observational errors in the field, all the head observations are corrupted with a Gaussian noise of zero mean and 0.1 m standard deviation.

In this example, the aquifer's transmissivity is selected as a parameter to be identified. Therefore, the main objective is to identify the transmissivity zone structure through inverse modeling approach described in the previous section. The true transmissivity field of the aquifer system is shown in Figure 3. Note that in this field, the largest and smallest transmissivity values are about 595 m^2/day and 33 m^2/day.

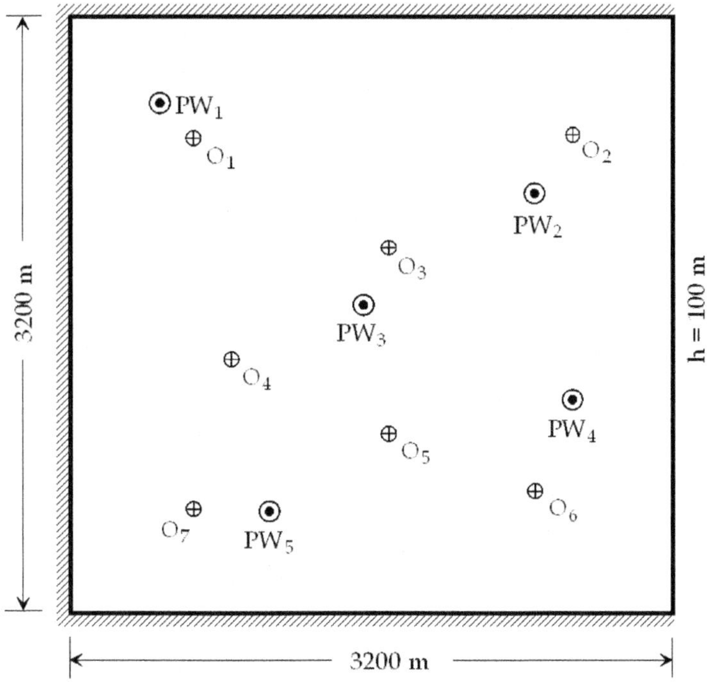

Fig. 2. The hypothetical aquifer model

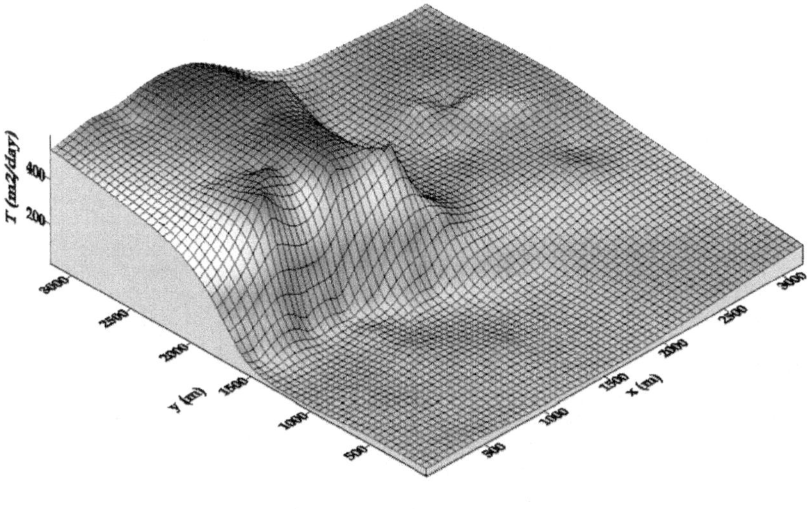

Fig. 3. True transmissivity field to be identified

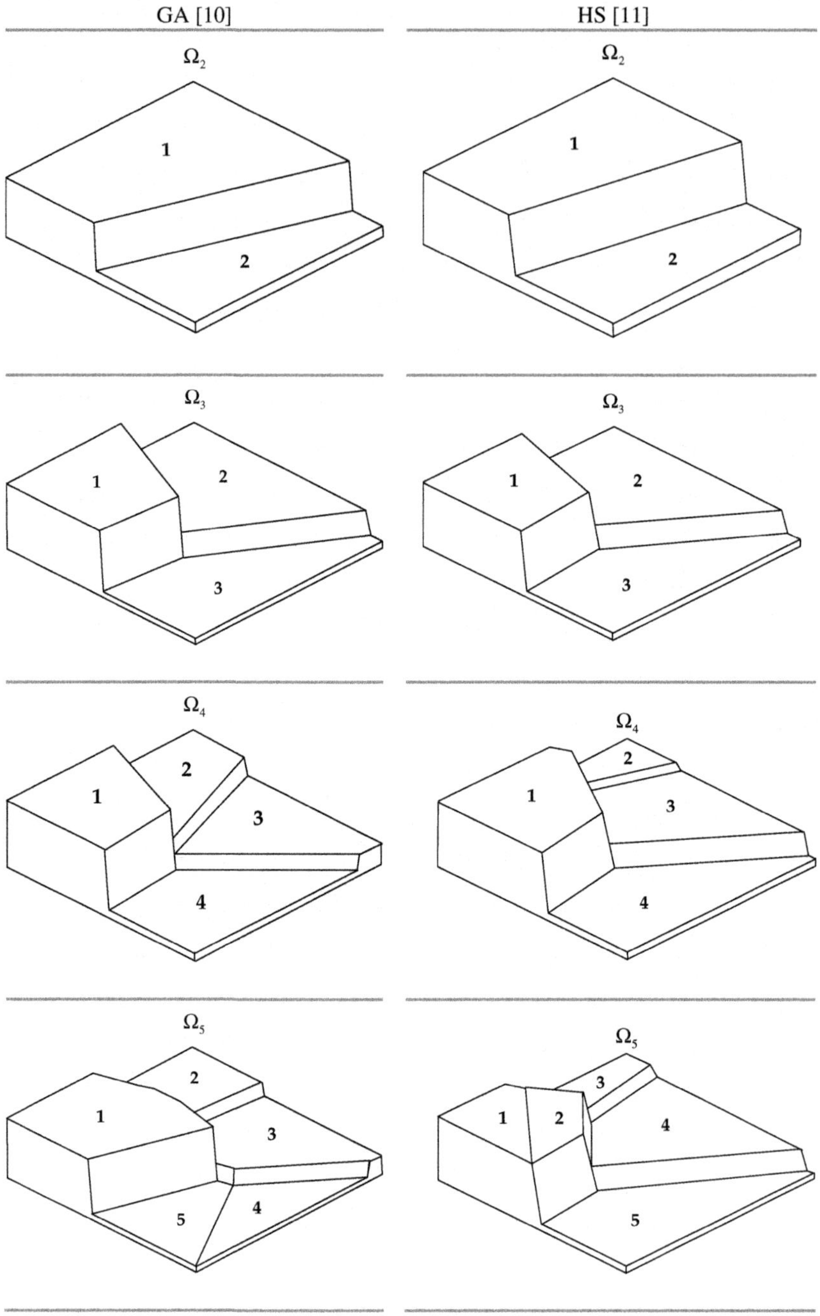

Fig. 4. Comparison of the identified zone structures

3.2 Comparison of the Identified Results

Table 1 and 2 list the results of GA [10] and HS [11]. After comparing both tables, it can be seen that RE values decrease with the increase of the zone numbers. However, the improvement of the RE is not significant after Ω_3. As can be seen, HS finds better results than GA (+ grid search + BFGS) in terms of the final RE values. Also, HS requires fewer simulation runs than GA. Although the obtained transmissivities are nearly same in Ω_1, they are different at the others. The reason of this is related to the differences of the identified zone structures.

Table 1. Identified results for different zone structures using HS

Structures	RE	# of Simulations	T_1	T_2	T_3	T_4	T_5
Ω_1	275.95	3,690	208	–	–	–	–
Ω_2	60.46	12,732	281	56	–	–	–
Ω_3	2.50	23,486	585	233	58	–	–
Ω_4	2.33	29,370	499	259	218	57	–
Ω_5	2.31	33,363	541	557	284	214	56

Table 2. Identified results for different zone structures using GA

Structures	RE	# of Simulations	T_1	T_2	T_3	T_4	T_5
Ω_1	280.11	10,000	207	–	–	–	–
Ω_2	71.60	15,000	253	46	–	–	–
Ω_3	3.55	25,000	517	228	52	–	–
Ω_4	2.62	40,000	501	299	164	54	–
Ω_5	2.61	50,000	520	279	172	48	51

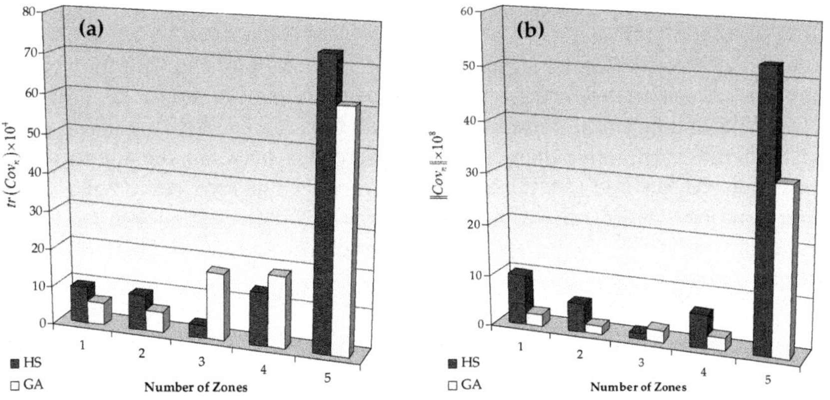

(a) (b)

Fig. 5. Variation of the PU values for different zone structures; (a): for the trace of the covariance matrix; (b): for the norm of the covariance matrix

The comparison of the identified zone structures for HS and GA solutions can be seen in Figure 4. As can be seen from Figure 4, all the zone structures well capture the true transmissivity field for both HS and GA solutions. Note that Tsai et al. [10] used the VT as a zonation technique in the GA solution whereas Ayvaz [11] used the FCM in the HS solution.

As can be seen from Figure 4, both HS and GA can effectively identify the zone structures of the transmissivity field given in Figure 3. However, the important issue at that point is to determine the optimum zone structure which best represents the true transmissivity field. Note that, when the number of zones increases, the number of the decision variables also increases. For instance, while there are 6 decision variables for Ω_2 solution (x and y coordinates of the basis points and embedded transmissivity values for each zone), the number of them is 15 for Ω_5 solution. As indicated in the previous section, when the number of decision variables increases, the reliability of them usually decreases. In other words, uncertainty of the identified parameters usually increases. Hence, uncertainty of the identified parameters is also checked during the simulation. Figure 5 shows the trace and the norm of the covariance matrix of the identified parameters which are the two common measures of calculating the PU. As can be seen from Figure 5, the change in the PU values is not significant for Ω_1 - Ω_4 solutions. However, there is a sudden jump in the Ω_5 solution. This means the reliability of the identified parameters decreases at that zone structure and the use of Ω_5 as an optimum zone structure is not appropriate in both HS and GA solutions. Therefore, use of Ω_4 is appropriate as an identified zone structure for both HS and GA.

4 Conclusions

This chapter reviewed the application of the HS algorithm to the solution of parameter structure identification problems in groundwater modeling. In order to evaluate the applicability of the HS algorithm to finding the global optimum solution for a parameter structure identification problem, the performance of HS was compared with a GA based hybrid solution model given in literature. The comparison showed that HS gives better results than GA in terms of the final objective function values and requires less simulation runs.

Although the identified zone structures of HS well capture the true transmissivity field for all the zone structures, there are some differences between the identified zone structures of HS and GA. The reason for this may be associated with a given parameterization schemes. Although there are the differences between the identified zone structures, both the HS and GA based solution models finds the same zone structure as an optimum one. The results of the comparison also indicate that HS can be effectively used for identifying the parameter zone structures and the associated parameter values for the solution of the inverse problem.

References

1. Ayvaz, M.T., Karahan, H.: A simulation/optimization model for the identification of unknown groundwater well locations and pumping rates. Journal of Hydrology 357, 76–92 (2008)
2. Theodossiou, N.: Combined use of simulation and optimization models in aquifer management: A case study. Transactions on Ecology and the Environment 7, 73–80 (1994)

3. Ayvaz, M.T., Karahan, H., Aral, M.M.: Aquifer parameter and zone structure estimation using kernel-based fuzzy c-means clustering and genetic algorithm. Journal of Hydrology 343, 240–253 (2007)
4. McKinney, D.C., Lin, M.-D.: Genetic algorithm solution of groundwater management models. Water Resources Research 30, 1897–1906 (1994)
5. Aral, M.M., Guan, J., Maslia, M.L.: Identification of contaminant source location and release history in aquifers. Journal of Hydrologic Engineering 6, 225–234 (2001)
6. Guan, J., Aral, M.M.: Optimal remediation with well locations and pumping rates selected as continuous decision variables. Journal of Hydrology 221, 20–42 (1999)
7. Willis, R.L., Yeh, W.W.G.: Groundwater systems planning and management. Prentice-Hall, Englewood Cliffs (1987)
8. Sun, N.-Z.: Inverse problems in groundwater modeling. Kluwer Academic Publishers, Dordrecht (1994)
9. Geem, Z.W., Kim, J.H., Loganathan, G.V.: A new heuristic optimization algorithm: Harmony search. Simulation 76, 60–68 (2001)
10. Tsai, F.T.-C., Sun, N.Z., Yeh, W.W.G.: A combinatorial optimization scheme for parameter structure identification in ground-water modeling. Groundwater 41, 156–169 (2003)
11. Ayvaz, M.T.: Simultaneous determination of aquifer parameters and zone structures with fuzzy c-means clustering and meta-heuristic harmony search algorithm. Advances in Water Resources 30, 2326–2338 (2007)
12. Jon Anderson's Web Site (2008) (accessed October 29, 2008), `http://dreams-in-color.com/sketch.html`
13. Wikipedia, Harmony Search (2002) (accessed October 29, 2008), `http://en.wikipedia.org/wiki/Harmony_search`
14. Geem, Z.W., Choi, J.-Y.: Music composition using harmony search algorithm. In: Giacobini, M. (ed.) EvoWorkshops 2007. LNCS, vol. 4448, pp. 593–600. Springer, Heidelberg (2007)
15. Geem, Z.W.: Harmony search algorithm for solving sudoku. In: Apolloni, B., Howlett, R.J., Jain, L. (eds.) KES 2007, Part I. LNCS (LNAI), vol. 4692, pp. 371–378. Springer, Heidelberg (2007)
16. Saka, M.P.: Optimum design of steel sway frames to BS5950 using harmony search algorithm. Journal of Constructional Steel Research 65, 36–43 (2009)
17. Geem, Z.W.: Optimal cost design of water distribution networks using harmony search. Engineering Optimization 38, 259–277 (2006)
18. Geem, Z.W., Lee, K.S., Park, Y.: Application of harmony search to vehicle routing. American Journal of Applied Sciences 2, 1552–1557 (2005)
19. Geem, Z.W.: Optimal scheduling of multiple dam system using harmony search algorithm. In: Sandoval, F., Prieto, A.G., Cabestany, J., Graña, M. (eds.) IWANN 2007. LNCS, vol. 4507, pp. 316–323. Springer, Heidelberg (2007)
20. Ayvaz, M.T.: Application of meta-heuristic harmony search optimization algorithm for parameter structure identification in groundwater modeling. In: Proceedings of the 8th International Congress on Advances in Civil Engineering, pp. 559–566 (2008)
21. Cheng, Y.M., Li, L., Lansivaara, T., Chi, S.C., Sun, Y.J.: An improved harmony search minimization algorithm using different slip surface generation methods for slope stability analysis. Engineering Optimization 40, 95–115 (2008)
22. Vasebi, A., Fesanghary, M., Bathaee, S.M.T.: Combined heat and power economic dispatch by harmony search algorithm. International Journal of Electrical Power & Energy Systems 29, 713–719 (2007)

23. Ceylan, H., Ceylan, H., Haldenbilen, S., Baskan, O.: Transport energy modeling with meta-heuristic harmony search algorithm, an application to Turkey. Energy Policy 36, 2527–2535 (2008)
24. Harmony Search Algorithm Web Site (2008) (accessed October 29, 2008), `http://www.hydroteq.com/`
25. Yeh, W.W.G.: Review of parameter-identification procedures in groundwater hydrology - the inverse problem. Water Resources Research 22, 95–108 (1986)
26. Tung, C.P., Chou, C.A.: Pattern classification using tabu search to identify the spatial distribution of groundwater pumping. Hydrogeology Journal 12, 488–496 (2004)
27. Sun, N.Z., Yang, S.L., Yeh, W.W.G.: A proposed stepwise regression method for model structure identification. Water Resources Research 34, 2561–2572 (1998)
28. Sun, N.Z., Yeh, W.W.G.: Identification of parameter structure in groundwater inverse problem. Water Resources Research 21, 869–883 (1985)
29. Eppstein, M.J., Dougherty, D.E.: Simultaneous estimation of transmissivity values and zonation. Water Resources Research 32, 3321–3336 (1996)
30. Sun, N.Z., Jeng, M.C., Yeh, W.W.G.: Model structure identification: the generalized inverse problem. In: Proceedings of the International Conference on Advances in Groundwater Hydrology, pp. 130–134 (1997)
31. Tung, C.P., Tang, C.C., Lin, Y.P.: Improving groundwater-flow modeling using optimal zoning methods. Environmental Geology 44, 627–638 (2003)
32. Tung, C.P., Tan, C.C.: An optimal procedure for identifying parameter structure and application to a confined aquifer. Environmental Geology 47, 1062–1071 (2005)
33. Voronoi, G.F.: Nouvelles applications des paramètres continus à la théorie de formes quadratiques. Journal für die reine und angewandte Mathematik 134, 198–287 (1908)
34. Bezdek, J.C.: Pattern recognition with fuzzy objective function algorithms. Plenum Press, New York (1981)
35. Yeh, W.W.G., Yoon, Y.S.: Aquifer parameter identification with optimum dimension in parameterization. Water Resources Research 17, 664–672 (1981)

Modified Harmony Methods for Slope Stability Problems

Yung-Ming Cheng

Department of Civil and Structural Engineering, Hong Kong Polytechnic University, Hong Kong
ceymchen@polyu.edu.hk

Abstract. In many practical slope stability problems, a thin layer of soft material with very low shear strength parameters may exist. Such soft and thin soil layer has caused many large scale slope failures in Hong Kong and many other places. Determining the critical failure surface with the smallest factor of safety for such a problem is extremely difficult. This chapter proposes a domain transformation technique, which is coupled with two new versions of improved harmony search algorithms to ensure both the efficiency and effectiveness of optimization analysis for large scale difficulty problems. The improved harmony search algorithms differ from the original harmony search algorithm in that: (1) the harmonies are rearranged into several pairs and the better pairs are used to develop several new harmonies; (2) different probabilities are assigned to different harmonies. The robustness of the newly proposed methods is demonstrated by using some difficult slope stability problems, and the improvement in the efficiency of the new algorithms is examined with respect to some large scale slope stability problems in China.

Keywords: Slope Stability, Factor of Safety, Modified Harmony Search, Objective Function, Global Minimum.

1 Introduction

For the stability analysis of slope by the limit equilibrium method, the minimum factor of safety is the basis for the design of slope stabilization measures. It is crucial that a good result should be determined because failures of slopes have caused significant loss of lives and properties in Hong Kong and many other cities. The evaluation of the minimum factor of safety of a slope involves the generation of trial slip surfaces, evaluation of the factors of safety for the trial slip surfaces and determination of the critical slip surface with the lowest factor of safety. By nature, the slope stability problem is a difficult non-convex non-polynomial type global optimization problem whose solution is difficult, and the objective function may also be discontinuous at different locations within the solution domain. Many engineers are facing this problem in their routine design works, and many of them are using the simple trial and error approach with the support of experience. Cheng et al. [1] has demonstrated that the use of the modern optimization method can greatly improve the results of analysis as compared with the trial and error approaches used in some practical problems in Hong Kong. At present, most of the engineers are still using the classical simplex method or Monte Carlo simulation method in the optimization analysis (adopted in some commercial software) which appears to function well for simple problems, but some engineers are experiencing limitations in use of these methods in some difficult problems.

Z.W. Geem (Ed.): Music-Inspired Harmony Search Algorithm, SCI 191, pp. 141–162.
springerlink.com

Optimization search for the non-circular failure surface (large number of control variables) has been considered by various classical methods (mainly gradient type method) by Baker and Gaber [2], Nguyen [3], Celestino and Duncan [4], Arai and Tagyo [5], Baker [6], Yamagami and Jiang [7], Chen and Shao [8]. These classical methods are all limited by the presence of local minimum as the local minimum close to the initial trial will be obtained using these methods.

In view of the limitations of the classical optimization methods, the current approach is the adoption of the stochastic global optimization methods for the present problem. These algorithms usually find a solution which is close to the best one with good efficiency. Greco [9] and Malkawi et al. [10] adopted Monte Carlo technique for searching the critical slip surface with success for some cases, but there is no precision control on the accuracy of the global minimum. Zolfaghari et al. [11] adopted the genetic algorithm while Bolton et al. [12] used the leap-frog optimization technique to evaluate the minimum factor of safety. The above methods are based on the use of static bounds to the control variables where the solution domain for each control variable is fixed and pre-determined by engineering experience. Cheng [13, 14] has developed a procedure, which transformed the various constraints and requirements of a kinematically acceptable failure mechanism to the evaluation of the upper and lower bounds of the control variables and employed a simulated annealing algorithm to determine the critical slip surface. The control variables are defined with dynamic domains where the bounds are controlled by the requirement of a kinematically acceptable failure mechanism and are changing during the solution. Through such approach, there is no need to define the fixed solution domain for each control variable by engineering experience. The approach by Cheng [13] is actually equivalent to the enforcement of convex shape by limiting the upper and lower bounds of the control variables sequentially, and this approach should be applicable to other disciplines which require a convex shape (but not necessarily a convex function). On the other hand, by removing some of the constraints of the lower bounds, the approach can also be adopted to non-convex shapes.

Among the modern stochastic global optimization methods that have evolved in recent years, there have been applications in geotechnical engineering using the genetic algorithm, simulated annealing, tabu search, ant-colony optimization, particle swarm optimization (PSO), leap-frog algorithm, and harmony search (HM) [11-16]. Cheng et al. [16] have carried out a detailed comparisons between six major types of stochastic global optimization methods, and the sensitivity of these methods under different optimization parameters have been investigated. Later, fish swarm algorithm and a modified harmony search algorithm were also applied to the slope stability problem [17, 18].

Some of the special problems for a robust and versatile optimization algorithm for slope stability analysis which may not be present in common optimization problems include:

(1) Discontinuity in some solution domain because the generated slip surface may be kinematically inadmissible or no factor of safety can be determined (no converged solution).

(2) Presence of many local minima which has been demonstrated by Chen and Shao [8].

(3) Soil parameters, ground condition, and external loads can vary over a wide range so that the choices of the optimization parameters are difficult to be established.

(4) The number of control variables can be very large and solution time can be long if a refined analysis is required or the size of the solution domain is large.

In view of the limitations of the classical optimization methods in slope stability analysis, several stochastic global optimization methods were developed mainly from pattern recognition, electronic production or control engineering. Also, the signal processing system is under consideration by the author. Some of the new global optimization methods such as that based on the genetic algorithm [11, 12] and the simulated annealing method [13, 16] have been used successfully in many problems, but the author has shown that these methods are less efficient when the number of control variables is large. For the harmony search method, the author has found that the original harmony search method can be trapped easily by the presence of local minima if the number of control variables is large and the objective function is discontinuous within the solution domain. Because of the limitation in the basic harmony search method, two improved harmony search algorithms (NHS1 and NHS2) are proposed in this chapter. NHS1 and NHS2 have been found to be effective and efficient for general problems, and four relatively difficult examples were investigated to demonstrate the effectiveness of the proposed method. To deal with the case of a narrow zone where there is a rapid change in the design parameters, the author proposes the use of a domain transformation method (equivalent to a weighted random number) which is coupled with the present modified harmony search method, and this approach can be useful in geotechnical as well as other types of problems.

2 Generation of Trial Failure Surfaces

In global optimization analysis for a slope stability problem, there are two major issues to be considered: generation of failure surfaces and the determination of minimum factor of safety. The slip surface generation method used in this chapter [15, 16] is applicable to both concave and convex slip surface for slope stability problem. The procedures for generating arbitrary slip surface is shown in Figure 1, where the ground surface is represented by $y=y_g(x)$, bedrock is represented by $y=R(x)$. The trial slip surface in the figure is composed of $n+1$ vertices with coordinates of (x_i, y_i), in which i varies between 1 and $n+1$, thus there will be n vertical slices with slice angle α_i, in which i varies between 1 and n. The vertices V_1 and V_{n+1} are the exit and entrance points of the slip surface respectively, and the lower and upper bounds for these two vertices can be prescribed by the engineers easily. Usually, the x- and y-ordinates of these vertices are taken as the variables. The x-coordinates of V_2 to V_n can be obtained by even spacing as given by Eq. 1 while the upper and lower bounds to the y-ordinates $y_{i,max}$ and $y_{i,min}$ can be calculated by using Eq. 2.

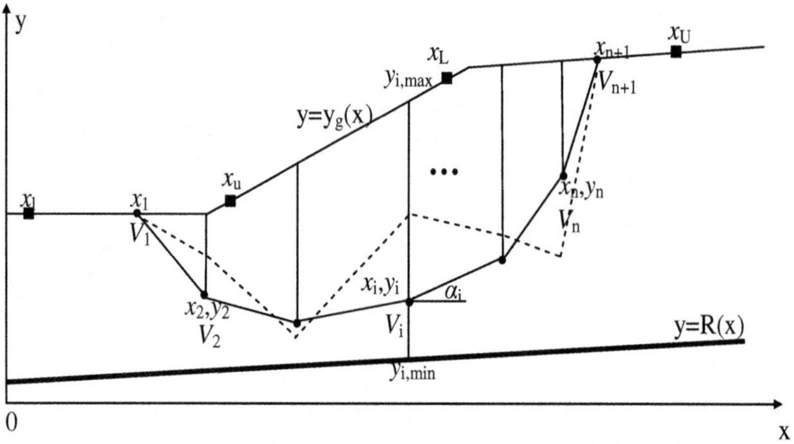

Fig. 1. A typical slip surface obtained by conventional procedure

$$x_i = x_1 + \frac{x_{n+1} - x_1}{n} \times (i-1), i = 2, ..., n \tag{1}$$

$$\begin{cases} y_{i,\max} = y_g(x_i) \\ y_{i,\min} = R(x_i) \end{cases}, i = 2, ..., n \tag{2}$$

The unacceptable slip surface as shown in Figure 1 (dash line) is obtained by random numbers. In order to render it to be kinematical, a correcting procedure is proposed.

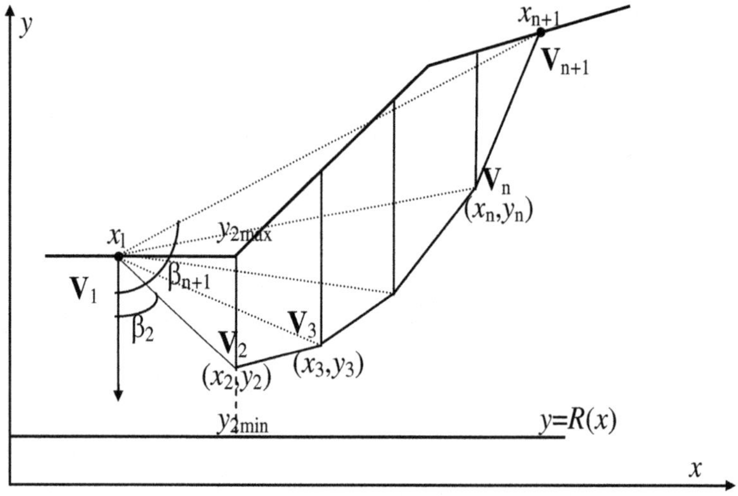

Fig. 2. Procedure for generating trial slip surface

From Figure 2, a trial slip surface consists of n+1 vertices, each of which is identified by the x- and y- coordinates of x_i,y_i, where i ranges from 1 to n+1. The exit point and entry points (V_1 and V_{n+1}) are controlled by using the upper and lower limits given by x_l,x_u and x_L,x_U. The x-coordinates of vertices from 2 to n are calculated by using the following Eq. 3.

$$x_i = x_1 + \frac{x_{n+1} - x_1}{n} \times (i - 1 + \zeta_i), i = 2, ..., n \tag{3}$$

where ζ_i is a variable used to determine the x-coordinates of V_i and is subjected to -0.5 < ζ_i < 0.5. Given the value of x_2, we can calculate y_{2max} and y_{2min} using Eq. 2. A variable χ_2 within the range of 0 to 1 is introduced to represent the y-coordinates of V_2. y_2 can be obtained using Eq. 4.

$$y_2 = y_{2\min} + \left(y_{2\max} - y_{2\min}\right) \times \chi_2 \tag{4}$$

Two lines are obtained by connecting V_1 and V_{n+1}, V_1 and V_2, respectively. The angles of the lines from the downward direction V_1 are denoted as β_2,..., β_{n+1} respectively. n-2 values are randomly generated within the range of β_2 to β_{n+1} using n-2 values of χ_3,..., χ_n in the range of 0 to 1.0 as given by Eq. 5 and they are sorted by ascending order and are related to β_3 to β_n, respectively. The y-coordinates of V_3 to V_n are determined by Eq. 6.

$$\beta_i = \chi_i \times \left(\beta_{n+1} - \beta_2\right) \tag{5}$$

$$y_i = y_1 + \tan\left(\beta_i - \pi/2\right) \times \left(x_i - x_1\right) \tag{6}$$

where i = 3,..., n.

Each trial slip surface obtained by this procedure can be represented mathematically by the vector $\mathbf{V} = \left(x_1, \zeta_2, \chi_2; ..., \zeta_n, \chi_n; x_{n+1}\right)$, in which $2n$ variables are used to determine the slip surface. The lower bound and upper bound to each element in V are denoted as l_i and u_i, respectively, where i ranges from 1 to $2n$. The proposed procedure cannot guarantee that all the generated trial slip surfaces are admissible, so the treatment as shown in Figure 3 will be necessary.

For the generated failure surface as shown in Figure 3, CD can be considered as unacceptable as this will be a restraint to the slope failure. CD can be kept for analysis if necessary but it will not be the critical solution in most cases so that a more efficient search is to correct CD to CD' so that the failure mechanism is acceptable. D' is determined by extrapolating BC to D', where D' and D have the same x-ordinate. Vertices D should be moved to D', and the same check should be performed for every adjacent three vertices until the whole failure surface is acceptable. If D is higher than D', CD will be acceptable and no modification is required. In view of the

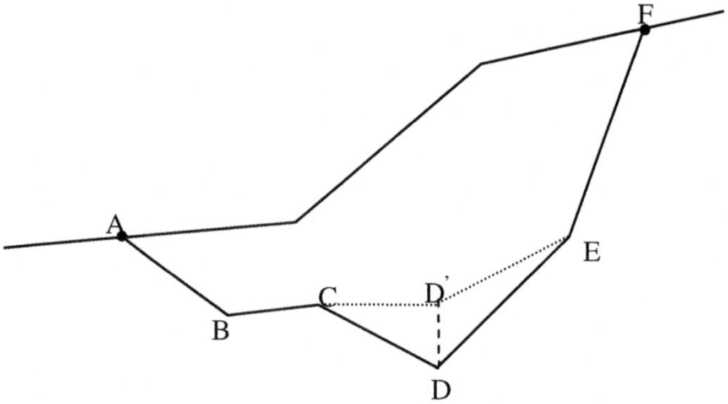

Fig. 3. Correction procedures

above process, the optimization problem associated with the location of the critical slip surface for the present formulation is outline as Eq. 7.

$$\begin{aligned} &\min \quad f(\mathbf{v}) \\ &s.t. \quad x_l \le x_1 \le x_u \qquad x_L \le x_{n+1} \le x_U \qquad -0.5 < \zeta_i < 0.5 \qquad i = 2,\dots,n \\ &\beta_2 < \beta_i < \beta_{n+1} \Longleftrightarrow 0 < \chi_i < 1 \qquad i = 3,\dots,n \\ &y_{2\min} \le y_2 \le y_{2\max} \Longleftrightarrow \quad 0 < \chi_2 < 1 \end{aligned} \tag{7}$$

It must be noticed that the trial slip surface of n+1 vertices is generated by using m = 2n variables.

3 Modified Harmony Search for Slope Stability Analysis

Harmony search algorithm was developed by Geem et al. [19, 20], which was conceptualized based on the musical process of searching for a perfect state of harmony. Musical performances seek to find pleasing harmony (a perfect state) as determined by an aesthetic standard, just as the optimization process seeks to find a global solution as determined by an objective function. The harmony in music is analogous to the optimization solution vector, and the musician's improvisations are analogous to local and global search schemes in optimization techniques. The HS algorithm does not require initial values for the decision variables. Furthermore, instead of a gradient search, the HS algorithm uses a stochastic random search that is based on the harmony memory considering rate *HMCR* and the pitch adjusting rate *PAR* so that derivative of the objective function is unnecessary during the analysis.

Harmony search algorithm (HS) is a population based search method. A harmony memory HM of size M (usually equals 2m) is used to generate a new harmony which is probably better than the optimum in the current harmony memory. The harmony memory consists of M harmonies (slip surfaces) and the M harmonies are usually generated randomly. Consider

$$HM = \{hm_1 \quad hm_2 \quad ,..., \quad hm_M\}, \; hm_i = (v_{i1}, v_{i2}, ..., v_{im}) \tag{8}$$

where each element of hm_i corresponds to that in vector V as described above. The most important procedure in HS is the generation of a new harmony named hm_{M+1}, and its solution procedures are outlined in Figure 4.

Harmony search algorithm has the following steps:

Step 1: Initialize the algorithm parameters: *HMCR, PAR, and M*, then randomly generate *M* harmonies (slip surfaces).

Step 2: Generate a new harmony (as shown in Figure 4). Here, pitch adjustment action is expressed as

$$\begin{cases} v_{M+1,i} = v'_{M+1,i} + (u_i - v'_{M+1,i}) \times rd & r_i > 0.5 \\ v_{M+1,i} = v'_{M+1,i} - (v'_{M+1,i} - l_i) \times rd & r_i \le 0.5 \end{cases} \tag{9}$$

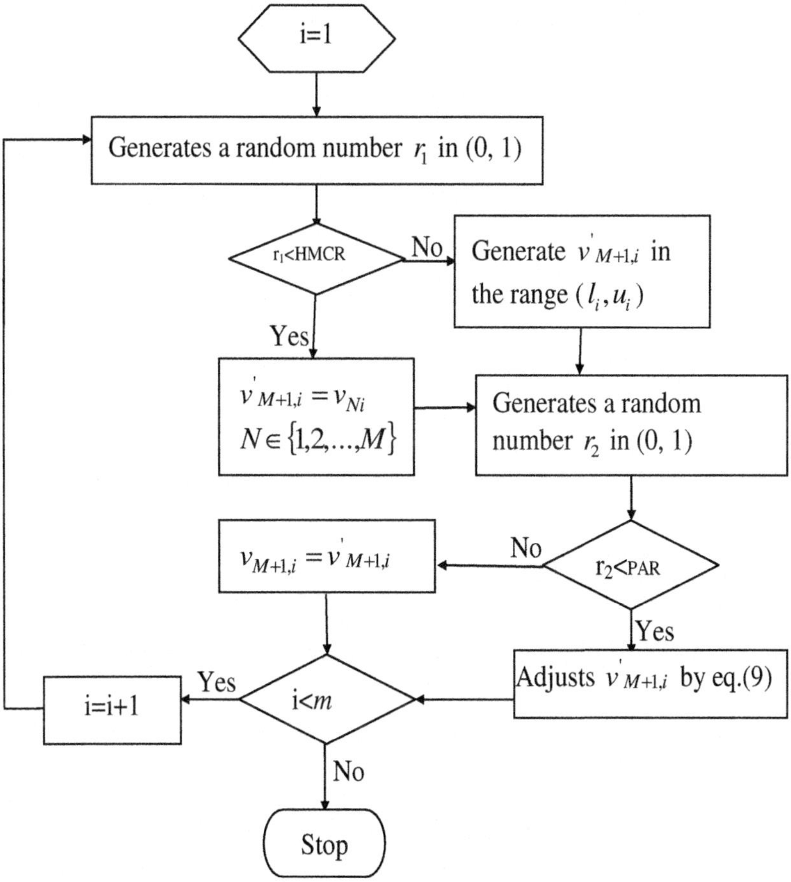

Fig. 4. The procedure for generating a new harmony

Step 3: Update the HM. If the new harmony is better than the worst harmony in the HM in terms of safety factor, the worst harmony is replaced with the new harmony.

Step 4: Repeat Steps 2 and 3 until the termination criterion is achieved.

Here, two random numbers *rd* and r_i in the range [0, 1] are used within different stages of the modified harmony search algorithm. In addition, the number of evaluations of objective function during the search, denoted as NEOF, can represent the computation time required by the optimization algorithm.

Based on many trials by the author, it is found that the original simple harmony (OHS) search algorithm works well for simple optimization problem with less than 25 control variables in slope stability problems where the objective function is not continuous over the solution domain. For more complicated problems with a large number of control variables, it is found that the results from the original harmony search algorithm may be unsatisfactory for some difficult problems as shown in a later section. The author has developed two modified harmony search algorithms for such difficult cases which differ from the original method in two aspects. The first difference is the probability of each harmony. The better the objective function value of one harmony, the more probable will the harmony be chosen for the generation of a new harmony. The second modification is that instead of one new harmony generated in the original method, several new harmonies (*Nhm*) are generated during each iteration step in the modified algorithm. In general, the modified harmony search method (NHS) is much more effective than the original harmony search method for large scale optimization but will be slightly less efficient for small scale problem. NHS is also more stable for small to large scale problems and is recommended for use.

In the original harmony search method (OHS), only one harmony is obtained from *M* harmonies in the current HM, and each harmony is used with the same probability. This is the main reason why the original harmony is highly efficient for small scale problems but can be trapped by the local minimum easily for large scale optimization problems. Actually, the *M* harmonies in HM can be classified into groups based on their objective function values, and the probability of the better harmony should be higher than the worse ones. In the new harmony search algorithm (NHS) proposed in this chapter, the harmonies are rearranged into *M*/2 pairs as illustrated in Figure 5.

In Figure 5, m_i (i = 1,…, *M*/2) is randomly chosen from *M*/2+1 to *M*. The pairs located at the front of HM have greater probability to generate new harmonies. We can introduce a parameter λ ($0<\lambda\leq 1$)) to decide which pair is used to generate new harmonies. An array *BB*() is used to represent the probabilities of all the pairs, where $BB(i) = \lambda(1-\lambda)^{i-1}, i = 1,\ldots, M/2$. The accumulated probability array $AC(\ )$ is calculated as $AC(i) = \sum_{j=1}^{i} BB(j)$. A random number r_a is obtained within the range of 0 to *AC*(*M*/2). If $AC(i-1) < r_a \leq AC(i)$, then the ith pair is used to generate one new harmony. The procedures for generating the new harmony by one given pair are outlined in Figue 6. Suppose i^{th} pair of harmonies, namely, hm_i and hm_k, are used to obtain a new harmony. A random number δ in the range of 0.2 to 0.8 (the lower and upper bounds to δ are based on the facts that the larger the value of δ, the more

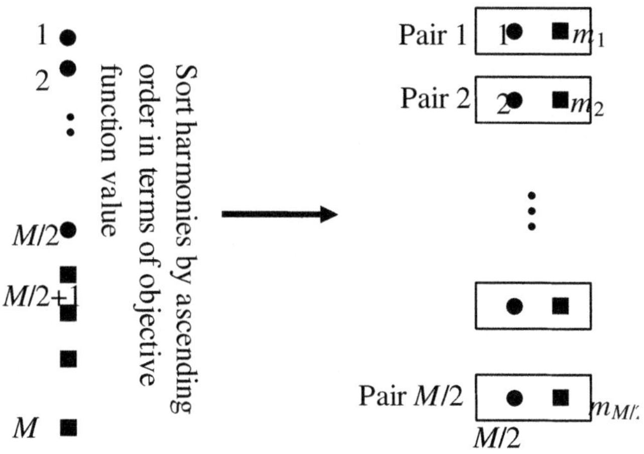

Fig. 5. Rearrangement for harmonies in HM

probable the better individuals in the current pair is chosen, and vice versa) is first randomly determined and the probabilities of hm_i and hm_k are ρ_i and ρ_k respectively using Eq. 10. The accumulated probabilities of hm_i and hm_k are ac_i and ac_k.

$$\begin{cases} \rho_i = \delta; \rho_k = \delta(1-\delta) \\ ac_i = \rho_i; ac_k = \rho_i + \rho_k \end{cases} \tag{10}$$

If one random number r_b within the range of 0 to ac_k is smaller than ac_i, then hm_i is used, and if r_b is higher than ac_i and smaller than ac_k, then hm_k is used. This procedure is called the choosing procedure.

Instead of only one new harmony is obtained in the original Harmony search algorithm, several new harmonies are generated in NHS. Two different versions of new harmony search methods are proposed here. In the first method NHS1, the iterative steps for NHS1 are as follows:

Step 1: Initialize the algorithm parameters: *HMCR*, *PAR*, *cp*, λ, *M* and randomly generate *M* harmonies (slip surfaces). When the counter $js = 0$, the optimal harmony is called hm_0, its objective function value is f_0, and its initial value is set to an arbitrary large value. For example, a value of 100 (much greater than normal factors of safety) is taken for slope stability analysis.

Step 2: As shown in Figure 5, the harmonies in HM are grouped in to *M*/2 pairs.

Step 3: Generate *M/2* random numbers rn_i, i=1,…, *M*/2 within the range of 0 to 1. If rn_i is lower than *cp*, a new harmony is obtained as illustrated in Figure 6, and altogether *D* new harmonies are obtained. One iteration step is finished and $js = js+1$.

Step 4: Evaluate *D* new harmonies and choose *M* harmonies into HM from *M* old harmonies and *D* new harmonies. The best harmony in HM is called hm_g and its corresponding value is f_g.

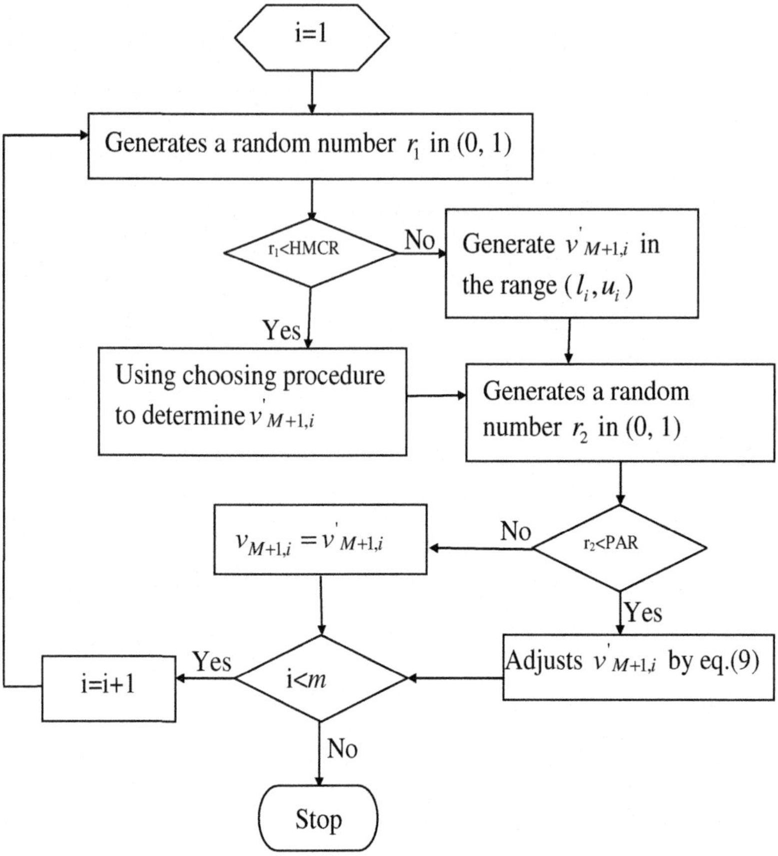

Fig. 6. The procedure for generating new harmony in NHS1

Step 5: If $js = N_1$, then check $|f_o - f_g| \leq \varepsilon$. If this relation is satisfied, the NHS will terminate, otherwise, set $N_1=N_1+N_2$, $hm_0=hm_g$, $f_0=f_g$ and NHS will go to Step 2. If $js<N_1$, NHS will go to Step 2.

It must be noted that $M/2$ random numbers and the parameter *cp* are used to determine the number of new harmonies to be generated within one iteration step in Step 3. An alternative procedure to determine the number of new harmonies is the use of parameter N_h which is the number of new harmonies to be obtained within one iteration step. The latter procedure is much more straightforward and is also easier to implement than the former one. In order to compare with the modified harmony search algorithm, in which the parameter N_h is adopted to determine the number of new harmonies within one iteration step, the same parameter N_h is also performed, namely, the latter procedure is adopted in this chapter. In the following example, the value of N_h is equal to 5, and N_1=500, N_2=200.

The second new harmony search method NHS2 differs from the original method in two aspects. The first difference is the probability of each harmony. Instead of the use of uniform probability in the original harmony search method, the better the objective function value of one harmony, the more probable will it be chosen for the generation of a new harmony. A parameter δ ($0<\delta\leq 1$) is introduced and all the harmonies in HM are sorted by ascending order, and a probability is assigned to each of them. For instance, *par(i)* means the probability to choose the ith harmony as follows:

$$par(i) = \delta \times (1-\delta)^{i-1} \tag{11}$$

where $i=1,2,\ldots, M$. From Eq. 11, it can be seen that the larger the value of δ, the more probable is the first harmony being chosen. An array $ST(i)(i=0, 1,\ldots, M)$ should be used to implement the above procedure for choosing the harmony.

$$ST(i) = \sum_{j=1}^{i} par(j) \tag{12}$$

where *ST(i)* represents the accumulating probability for i^{th} harmony. *ST(0)* is defined as 0.0 for the sake of implementation. A random number r_c is given form the range [0, *ST(M)*], and the kth harmony in HM is chosen if the following criterion is satisfied.

$$ST(k-1) < r_c \leq ST(k), k = 1,2,\ldots,M \tag{13}$$

The second modification in NHS2 is that rather than one new harmony, a certain number of new harmonies (*Nhm*) are generated during each iteration step in the modified algorithm. The utilization of HM is intuitively more exhaustive by generating several new harmonies than by generating only one harmony. In order to retain the structure of HM unchanged, the *M* harmonies with the lower objective functions (for the minimization optimization problem) from *M+Nhm* harmonies are included in the HM again and *Nhm* harmonies of higher objective functions are rejected.

Based on extensive tests on different types of problems, the author has found that the two improved harmony search methods usually perform better for more difficult problems (presence of multi local minima, large number of control variables). The number of evaluations and the results are better than the original harmony search method. On the other hand, for simple problems with smaller number of control variables, the original harmony search method is usually more efficient. The new methods require more solution parameters than the original harmony search method, but based on extensive internal tests by the author, it is found that the new methods are stable over wide range of problems and are more effective than the original method in most cases which will be illustrated by the following examples.

4 Weighted Random Number for Soft Band Problem

In natural slopes, a soft thin band of soil is sometimes found. The design parameters of this thin band of soil are much lower than those soils above and below the soft

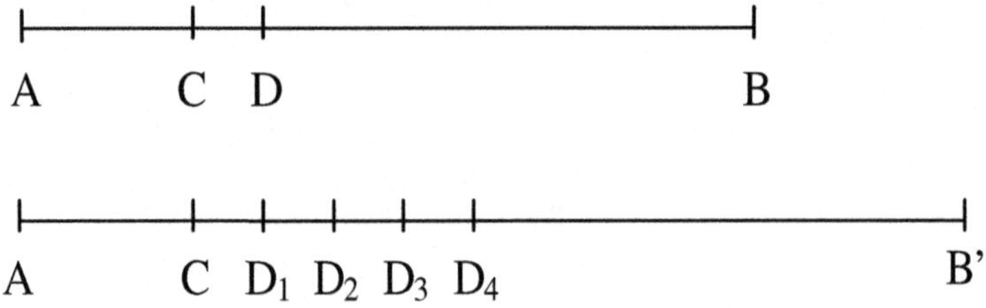

Fig. 7. Trick to generate a random number with greater probability for a special region

band. There is a much greater chance that the slip surface will fall within this soft band of soil. Since the location of this soft band is irregular in general, it is difficult to define a random number which will give greater probability to this narrow region at different x-ordinate. Thus, a special trick is adopted by the author so as to simulate a random number with special preference to a special region.

As shown in Figure 7, a classical random number is to be used within domain AB. In the figure, the actual domain for a control variable x_i (N+1> i >2) is represented by the segment AB with a soft band CD in between AB. For control variables x_j where i≠j, the location of the soft band CD and the solution bound AB for control variable x_i will be different from that for control variable x_j. For segment AB, several virtual domains with a width of CD for each domain are added adjacent to CD shown in Figure 7. The transformed domain AB' is used as the control domain of variable x_i. Every point generated within the virtual domain D1-D2, D2-D3, D3-D4 is mapped to the corresponding point in segment CD1. This technique is effectively equivalent to giving more chances to those control variables within the soft band. The weighting to the random number within the soft band zone can be controlled easily by the number of extra domains added as suggested in Figure 7.

5 Examples with New Harmony Search Methods

Based on the authors' internal study, the original harmony search method is found to be more efficient for small scale optimization problems. When the number of control variable is large and there is a small region in the solution domain where there is a major change in the soil parameters, it is found that the original harmony search method can be trapped by the presence of local minimum easily. For the present proposals, it is also found to work well for normal problems over wide range of soil parameters where there are no special geotechnical features. For problems with special geotechnical features, the robustness of the present optimization algorithm will be demonstrated by five relatively difficult examples where the precise location of the critical failure surface has a strong influence on the factor of safety and many global optimization methods may be trapped by the local minima easily.

5.1 Example 1

To illustrate the applicability of the proposed modified harmony search methods, four examples will be considered. Example 1 is a slope with four layers of soils shown in

Table 1. Geotechnical parameters for example 1

Layers	γ (kN/m^3)	c' (kPa)	ϕ' (degree)
1	19.0	15.0	20.0
2	19.0	17.0	21.0
3	19.0	5.00	10.0
4	19.0	35.0	28.0

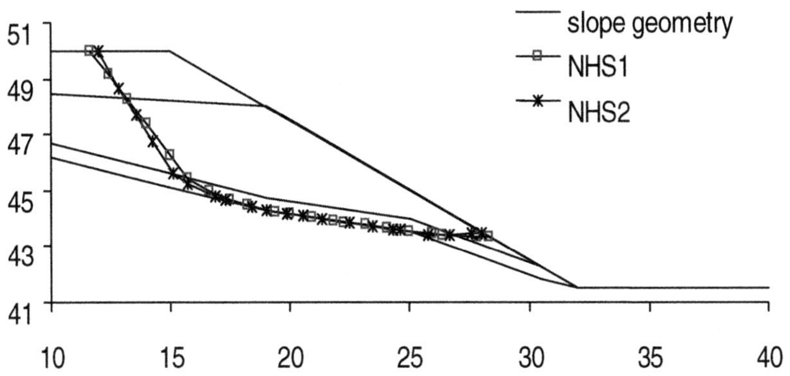

Fig. 8. Geotechnical profile and critical solution for example 1

Figure 8 while the soil parameters are given in Table 1 (γ, c' and ϕ' means the unit weight, effective cohesive strength and effective friction of soil). Soil layer 3 is a thin irregular weak layer and the Spencer method [21] is used to calculate the factor of safety. Since soil layer 3 is thin with poor soil parameters, many control variables should lie within this region which is not easily determined automatically from the OHS. This situation is particularly important when the number of control variables is large which will be demonstrated. The results shown in Figure 10 have clearly illustrated that the performance of the OHS is poor under this case, while the improved harmony method give much lower factor of safety with similar critical slip surface as shown in Figure 8.

All stochastic optimization methods rely on the use of some parameters which are difficult to be determined for general case. In general, these parameters are based on the statistics from large number of numerical tests. Due to the special problems in slope stability analysis as mentioned in previous section, the authors view that the performance of the algorithm with respect to the use of optimization parameters is very important and a statistical check has been carried out in this chapter. The randomly generated 50 series of parameters are used to check the robustness of the present algorithm. From Figure 9, the results corresponding to the 50 series of parameters are mainly located in the range of 1.28 to 1.33, with only one result higher than 1.40 and two results higher than 1.35. The average value of the 50 factors of safety is 1.31 and the average number of NEOF is 11,617.

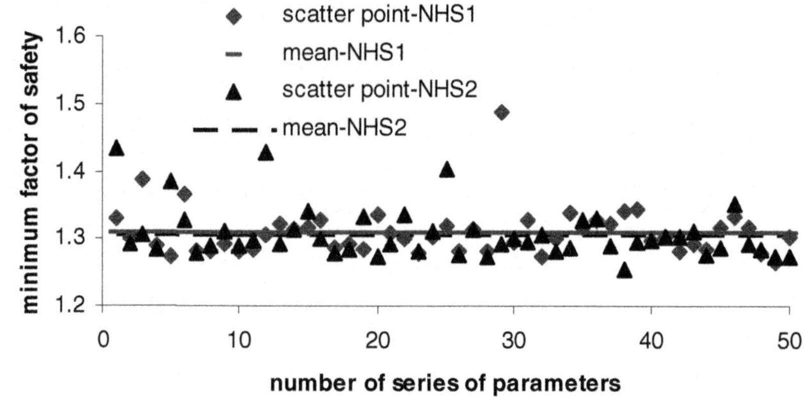

Fig. 9. Summary of the results for 50 series of parameters

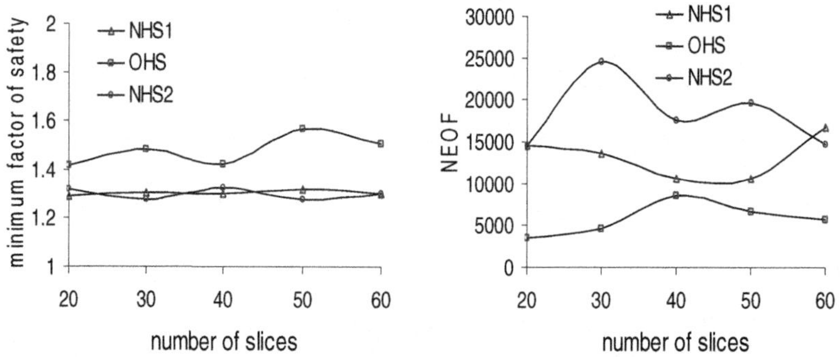

Fig. 10. Comparison of safety factor and NEOF for different number of control variables

If we assume the number of slice is equal to 20 (21 control variables) and four levels are set for three parameters (*HMCR, PAR,* λ), we obtain orthogonal results given in Table 2 (note that the parameter λ is not used in NHS2, so the number of total parameters is equal to 2).

Four levels for *HMCR* are (0.85, 0.9, 0.95, 1.0); (0.1, 0.15, 0.2, 0.25) for *PAR*; and (0.3, 0.4, 0.5, 0.6) for λ. The sensitivity of each parameter can be obtained through 16 tests. If the F value (Factorial Analysis of Variance [22]) of one parameter is larger than the critical value $F_{0.05}$ and is smaller than $F_{0.01}$, it implies that the calculated result is sensitive to this parameter. Otherwise, if the F value is smaller than $F_{0.05}$, it shows that the result is insensitive to this parameter. If the F value is larger than $F_{0.01}$, the result is hyper-sensitive to this parameter.

For NHS1, from the result in the 5th column in Table 2, the F values of the three parameters can be obtained as 0.78, 0.90 and 0.92, while $F_{0.05}$ = 4.8 and $F_{0.01}$ = 9.8. It can be concluded that the three parameters are insensitive to the analysis. Similarly, the F values of the two parameters (*HMCR, PAR*) used in NHS2 can be calculated as 0.47 and 0.59, compared with the standard value $F_{0.05}$ = 3.9 and $F_{0.01}$ = 7.0 (different

Table 2. Orthogonal test table by NHS1 and NHS2

	HMCR	*PAR*	λ	Results	
				NHS1	NHS2
1	1 (0.85)	1 (0.10)	1 (0.3)	1.3196	1.2938
2	1 (0.85)	2 (0.15)	2 (0.4)	1.3325	1.3276
3	1 (0.85)	3 (0.20)	3 (0.5)	1.3542	1.3321
4	1 (0.85)	4 (0.25)	4 (0.6)	1.3391	1.3413
5	2 (0.90)	1 (0.10)	2 (0.4)	1.3370	1.3079
6	2 (0.90)	2 (0.15)	1 (0.3)	1.3492	1.2917
7	2 (0.90)	3 (0.20)	4 (0.6)	1.3265	1.3298
8	2 (0.90)	4 (0.25)	3 (0.5)	1.3245	1.3057
9	3 (0.95)	1 (0.10)	3 (0.5)	1.3616	1.3235
10	3 (0.95)	2 (0.15)	4 (0.6)	1.2924	1.3223
11	3 (0.95)	3 (0.20)	1 (0.3)	1.3050	1.3134
12	3 (0.95)	4 (0.25)	2 (0.4)	1.3171	1.2783
13	4 (1.00)	1 (0.10)	4 (0.6)	1.3509	1.3327
14	4 (1.00)	2 (0.15)	3 (0.5)	1.3262	1.3298
15	4 (1.00)	3 (0.20)	2 (0.4)	1.2998	1.3144
16	4 (1.00)	4 (0.25)	1 (0.3)	1.2914	1.2903

number of parameters and levels result in different values of $F_{0.05}$ and $F_{0.01}$, and they can be determined easily from the existing tables), we can conclude that both parameters are insensitive to the NHS2 analysis. The results obtained by NHS1 and NHS2 are almost identical in this problem.

OHS is trapped by the local minima as shown in Figure 10 with relatively poor results for the safety factor, though the NEOF required by OHS is smaller than those required by NHS1 and NHS2. OHS located the safety factor within the range of 1.40 to 1.60 with different number of control variables, on the other hand, NHS1 and NHS2 find the minimum factor of safety cluster around 1.3 which is much better than the results by OHS. One special advantage of the harmony search is that the number of evaluations do not increase sharply with the increase in the number of control variables (number of slice), and this is a very important reason for the adoption of harmony search method for large scale slope stability analysis.

5.2 Example 2

Example 2, as shown in Figure 11, is a slope with three layers where an irregular weak layer is sandwiched between two strong layers. The geotechnical properties for layers 1 to 3 are friction angle equal to 20°, 10°, and 20°; cohesion equal to 28.73 kPa, 0.0 kPa, and 28.73 kPa; and unit weight 18.84 kN/m^3 for all three layers. The Spencer's method is adopted to determine the safety factor. As described above, the

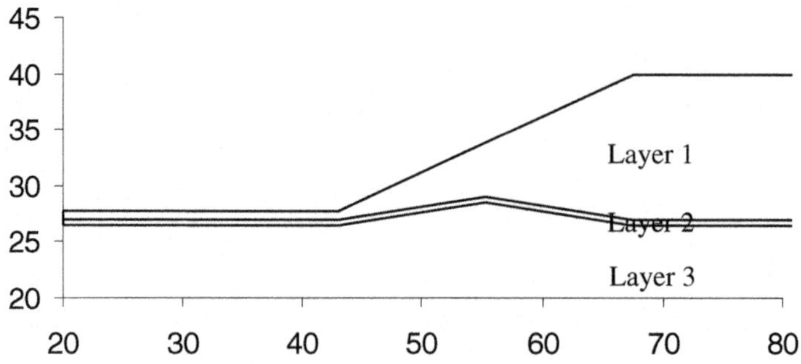

Fig. 11. Cross section of example 2

same 50 series of parameter sets randomly generated within their corresponding ranges are used to test the robustness of the proposed algorithm.

Assuming (x_l=30, x_u=35) and (x_L=70, x_U=75), the same orthogonal test table as used in example 1 is performed to investigate the sensitivities of parameters in the proposed algorithm. The obtained results are listed in Table 3.

Table 3. The orthogonal test table for example 2

	HMCR	*PAR*	λ	Results	
				NHS1	NHS2
1	1 (0.85)	1 (0.10)	1 (0.3)	1.4781	1.4578
2	1 (0.85)	2 (0.15)	2 (0.4)	1.4746	1.4379
3	1 (0.85)	3 (0.20)	3 (0.5)	1.4348	1.4849
4	1 (0.85)	4 (0.25)	4 (0.6)	1.4441	1.4707
5	2 (0.90)	1 (0.10)	2 (0.4)	1.3952	1.4034
6	2 (0.90)	2 (0.15)	1 (0.3)	1.4346	1.4041
7	2 (0.90)	3 (0.20)	4 (0.6)	1.4367	1.4684
8	2 (0.90)	4 (0.25)	3 (0.5)	1.4549	1.4092
9	3 (0.95)	1 (0.10)	3 (0.5)	1.3761	1.4690
10	3 (0.95)	2 (0.15)	4 (0.6)	1.3961	1.4154
11	3 (0.95)	3 (0.20)	1 (0.3)	1.4874	1.3912
12	3 (0.95)	4 (0.25)	2 (0.4)	1.4145	1.3789
13	4 (1.00)	1 (0.10)	4 (0.6)	1.4397	1.4071
14	4 (1.00)	2 (0.15)	3 (0.5)	1.4200	1.4052
15	4 (1.00)	3 (0.20)	2 (0.4)	1.4711	1.4654
16	4 (1.00)	4 (0.25)	1 (0.3)	1.4560	1.5514

The corresponding F values for the three parameters as easily obtained as follows: $F_{HR} = 1.48$, $F_{PR} = 1.12$, $F_{\lambda} = 1.68$.

For NHS2, the values become $F_{HR} = 1.15$, $F_{PR} = 0.57$. Compared with the standard F values, it can also be conclude that the parameters are insensitive to the optimization analysis.

The same values for the parameters used in OHS, NHS1, and NHS2 as used in example 1 are adopted to compare the results obtained by the three algorithms with the different number of slices. The results are summarized in Figure 12.

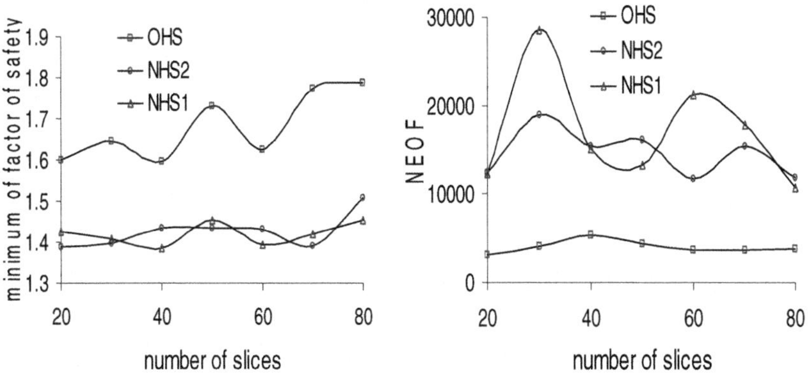

Fig. 12. Comparison of safety factor and NEOF with different number of slices

Results in Figure 12 are basically similar to those in Figure 10. OHS tends to locate a much higher safety factor than NHS1 and NHS2 for a given number of slices and the results fluctuates in the range of 1.60 to 1.80 with the number of slices varying from 20 to 80. On the other hand, the results by NHS1 and NHS2 are much more stable than OHS and the numbers of evaluation are also acceptable (most problems are finished within 5 minutes).

5.3 Example 3

Example 3, as shown in Figure 13, is based on a problem where an irregular weak layer is sandwiched between two strong layers. The geotechnical properties for soil layers 1 to 3 are φ' equal to 35°, 25° and 35°; c' equal to 20 kPa, 0 kPa and 10.0 kPa; and unit weight equals to 19.0 kN/m^3 for all the three soil layers.

The solution domain is assumed to be (x_l=3, x_u=8) and (x_L=20, x_U=28) for this example. As described above, the orthogonal analysis is firstly performed to investigate the sensitivity of the related parameters. The results obtained are listed in Table 4.

From the data listed in 5th column in Table 5, we can calculate the F values of three parameters which are equal to 7.95, 14.80, and 5.32 for NHS1. According to the comparison with the critical values $F_{0.05}$ and $F_{0.01}$, it can be concluded that *HMCR* and λ are sensitive and *PAR* is hyper-sensitive to the analysis. For NHS2, the corresponding F values are 5.82 and 2.28 and *HMCR* is sensitive while *PAR* is insensitive

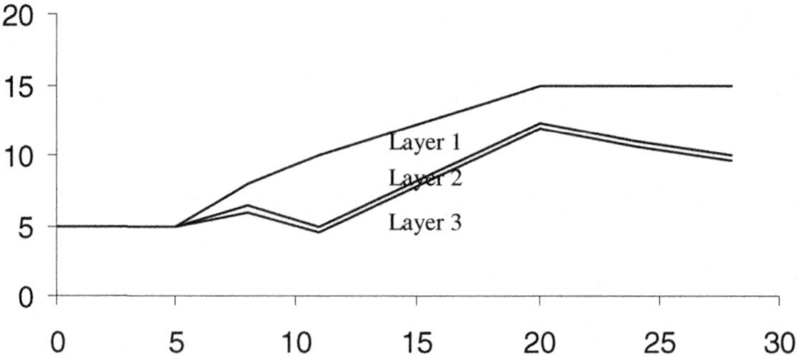

Fig. 13. Cross-section for example 3

Table 4. The orthogonal test table for example 3

	HMCR	*PAR*	λ	Results	
				NHS1	NHS2
1	1 (0.85)	1 (0.10)	1 (0.3)	1.4843	1.5419
2	1 (0.85)	2 (0.15)	2 (0.4)	1.5074	1.5462
3	1 (0.85)	3 (0.20)	3 (0.5)	1.5207	1.6225
4	1 (0.85)	4 (0.25)	4 (0.6)	1.5461	1.5227
5	2 (0.90)	1 (0.10)	2 (0.4)	1.5239	1.4814
6	2 (0.90)	2 (0.15)	1 (0.3)	1.4752	1.4872
7	2 (0.90)	3 (0.20)	4 (0.6)	1.5047	1.5345
8	2 (0.90)	4 (0.25)	3 (0.5)	1.5267	1.5215
9	3 (0.95)	1 (0.10)	3 (0.5)	1.4831	1.4681
10	3 (0.95)	2 (0.15)	4 (0.6)	1.4817	1.4923
11	3 (0.95)	3 (0.20)	1 (0.3)	1.4711	1.5107
12	3 (0.95)	4 (0.25)	2 (0.4)	1.5415	1.5385
13	4 (1.00)	1 (0.10)	4 (0.6)	1.4579	1.4779
14	4 (1.00)	2 (0.15)	3 (0.5)	1.4702	1.4970
15	4 (1.00)	3 (0.20)	2 (0.4)	1.4857	1.4870
16	4 (1.00)	4 (0.25)	1 (0.3)	1.5047	1.4828

to the analysis. It should be noted that the parameters used in the search algorithms are not sensitive for the simple examples, but they can be sensitive or hypersensitive for complicated examples. It is suggested that engineers should perform an orthogonal analysis for complicated problems. The authors have also found similar behavior for genetic algorithm and ant colony method. Actually, all stochastic global optimization methods will face similar problems in the choice of parameters, but the present modified proposals have greatly reduced this limitation which is critical for the original formulation.

The comparisons of results among OHS, NHS1 and NHS2, as seen in Figure14, show that the OHS results locate much higher than those by NHS1 and NHS2, although the NEOF required by OHS is smaller than those used by NHS1 and NHS2. This is mainly because, after N_2 iterations, there is no improvement on the safety factor obtained so far, which is called precocity. On the other hand, the NHS1 and NHS2 can improve the results found so far within N_2 iterations, so the NEOF used by them is always larger than that used by OHS. The largest one is not more than 30,000 and this is acceptable as the solution time is less than 5 minutes.

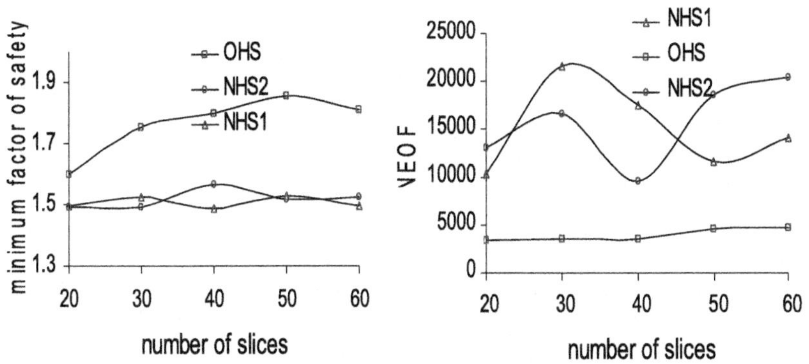

Fig. 14. Comparison of safety factor and NEOF with different number of slices

5.4 Example 4

Example 4, as shown in Figure 15, is an interesting problem which is worth consideration. The second layer of soil is only 1 mm thick so that it cannot be seen from the

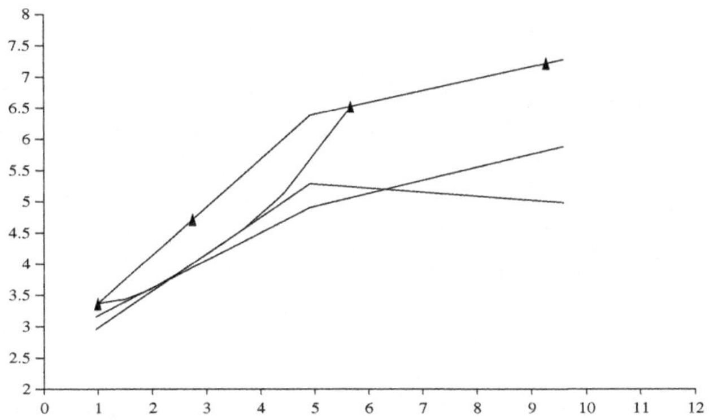

Fig. 15. Failure surface without domain transformation method (safety factor = 1.486)

figure. Since the soil properties of layer 2 are extremely low but this layer is also extremely thin, the majority part of the critical failure surface should lie within this zone. If the modified harmony search method is used without the domain transformation method as shown in Figure 7, a minimum safety factor of 1.486 is actually obtained as shown in Figure 15. On the other hand, if the domain transformation method is used, the minimum safety factor will be 1.037 which is shown in Figure 16. The difference in the results is extremely significant and this problem must be carefully considered. The present problem has a very thin domain where there is a sudden and major change of soil properties. If there is no special treatment to this special zone, no method can determine the minimum safety factor automatically.

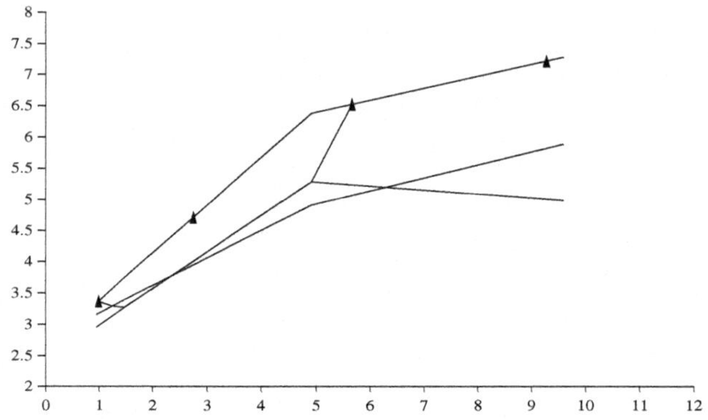

Fig. 16. Failure surface with domain transformation method (factor of safety = 1.037)

6 Conclusions

The original harmony search algorithm has been found to be trapped by the local minima for some large slope stability problem. The algorithm improvises only one new harmony in each iteration and uses equal probability for choosing the harmonies stored in HM. In views of the efficiency of the harmony search method and its limitations, two modified harmony search algorithms are proposed by generating pairs of harmonies and using different probabilities for different harmonies.

The modified harmony algorithms are found to be highly effective and efficient for many difficult problems. From numerous internal tests, it appears that the modified harmony methods are more effective and stable over a wide range of problems as compared with the original harmony method, except when the problems under consideration are simple with relatively few control variables.

Stochastic algorithms are approximate and not accurate algorithms. These algorithms usually find a solution closest to the best one. However, they usually find it fast and easily. Every stochastic algorithm relies on the use of some parameters for analysis but there is no serious method in determining these kinds of parameters. It is found that the two proposed algorithms are relatively insensitive to the use of these parameters due to the special arrangement in generating pairs of harmonies. This property is highly beneficial in slope stability problem.

The efficiency of a global optimization algorithm is another important factor to be considered. A majority of the global optimization methods will also require tremendous evaluations for large scale optimization analysis. The author has tested the proposed algorithms to a maximum 160 control variables (sufficient for most of the slope stability analyses) and have found that the proposed algorithms are still efficient and have performed satisfactorily. It is true that the original harmony method is usually more efficient than the modified methods for simple and small problems, but the original harmony method suffers from the limitation of being trapped by the presence of local minimum for the large scale problems.

The modified harmony algorithms and the trick in generating a random number with weighting to different zones have been proved to be highly effective in geotechnical engineering. The present algorithms are however not limited to geotechnical problems, and such algorithms should be applicable to general optimization problems. For the testing of the efficiency and effectiveness of the modified harmony search method, and the readers can download the demo program [23].

References

1. Cheng, Y.M., Lansivaara, T., Siu, J.: Impact of convergence on slope stability analysis and design. Computers and Geotechnics 35(1), 105–115 (2008)
2. Baker, R., Garber, M.: Theoretical analysis of the stability of slopes. Geotechnique 28, 341–395 (1978)
3. Nguyen, V.U.: Determination of critical slope failure surfaces. Journal of Geotechnical Engineering, ASCE 111, 238–250 (1985)
4. Celestino, T.B., Duncan, J.M.: Simplified search for non-circular slip surface. In: Proceeding of 10th International Conference on Soil Mechanics and Foundation Engineering, pp. 391–394 (1981)
5. Arai, K., Tagyo, K.: Determination of noncircular slip surfaces giving the minimum factor of safety in slope stability analysis. Soils and Foundations 21, 43–51 (1985)
6. Baker, R.: Determination of the critical slip surface in slope stability computations. International Journal of Numerical and Analytical Methods in Geomechanics 4, 333–359 (1980)
7. Yamagami, T., Ueta, Y.: Search for noncircular slip surfaces by the Morgenstern-Price method. In: Proceeding of 6th International Conference on Numerical Methods in Geomechanics, pp. 1219–1223 (1988)
8. Chen, Z., Shao, C.: Evaluation of minimum factor of safety in slope stability analysis. Canadian Geotechnical Journal 25(4), 735–748 (1983)
9. Greco, V.R.: Efficient Monte Carlo technique for locating critical slip surface. Journal of Geotechnical Engineering, ASCE 122, 517–525 (1996)
10. Malkawi, A.I.H., Hassan, W.F., Sarma, S.K.: Global search method for locating general slip surface using Monte Carlo techniques. Journal of Geotechnical and Geoenvironmental Engineering 127, 688–698 (2001)
11. Zolfaghari, A.R., Heath, A.C., McCombie, P.F.: Simple genetic algorithm search for critical non-circular failure surface in slope stability analysis. Computers and Geotechnics 32, 139–152 (2005)
12. Bolton, H.P.J., Heymann, G., Groenwold, A.: Global search for critical failure surface in slope stability analysis. Engineering Optimization 35, 51–65 (2003)

13. Cheng, Y.M.: Locations of critical failure surface and some further studies on slope stability analysis. Computers and Geotechnics 30, 255–267 (2003)
14. Cheng, Y.M.: Global optimization analysis of slope stability by simulated annealing with dynamic bounds and Dirac function. Engineering Optimization 39(1), 17–32 (2007)
15. Cheng, Y.M., Li, L., Chi, S.C., Wei, W.B.: Particle swarm optimization algorithm for location of critical non-circular failure surface in two dimensional slope stability analysis. Computers and Geotechnics 34(2), 92–103 (2007)
16. Cheng, Y.M., Li, L., Chi, S.C.: Performance studies on six heuristic global optimization methods in the location of critical failure surface. Computers and Geotechnics 34, 462–484 (2007)
17. Cheng, Y.M., Li, L., Chi, S.C., Wei, W.B.: Determination of critical slip surface using artificial fish swarms algorithm. Journal of Geotechnical and Geoenvironmental Engineering, ASCE 134(2), 244–251 (2008)
18. Cheng, Y.M., Li, L., Lansivaara, T., Chi, S.C., Sun, Y.J.: Minimization of factor of safety using different slip surface generation methods and an improved harmony search minimization algorithm. Engineering Optimization 40, 95–115 (2008)
19. Geem, Z.W., Kim, J.H., Loganathan, G.V.: A new heuristic optimization algorithm: harmony search. Simulation 76, 60–68 (2001)
20. Lee, K.S., Geem, Z.W.: A new meta-heuristic algorithm for continuous engineering optimization: harmony search theory and practice. Computer Methods in Applied Mechanics and Engineering 194, 3902–3933 (2005)
21. Spencer, E.: A method of analysis of the stability of embankments assuming parallel interslice forces. Geotechnique 17, 11–26 (1967)
22. Wikipedia, Analysis of variance (2001) (accessed December 4, 2008), `http://en.wikipedia.org/wiki/Analysis_of_variance`
23. Slope 2000 (2004) (accessed December 4, 2008), `http://www.cse.polyu.edu.hk/~ceymcheng/`

Harmony Search Algorithm for Transport Energy Demand Modeling

Halim Ceylan and Huseyin Ceylan

Department of Civil Engineering, Pamukkale University, Denizli, Turkey
{halimc,hceylan}@pau.edu.tr

Abstract. The transport sector is one of the major consumers of energy production throughout the world. Thus, the estimation of medium and long-term energy consumption based on socio-economic and transport related indicators is a critical issue on a global scale. This chapter reviews the harmony search (HS) applications to transport energy modeling problems. The models reviewed in this chapter are in the form of linear, exponential and quadratic mathematical expressions, and they are applied to transportation sector energy consumption. Convergence behavior of the HS during the modeling is tested. Performance of each HS model is compared according to an absolute error between observed data and predicted data. Results showed the HS method can be applied to the transport energy modeling issues.

Keywords: Transport Energy Modeling, Harmony Search, Optimization.

1 Introduction

Energy is one of the most important inputs for the economic development of a country. The importance of energy in economic development has been recognized almost universally; Historical data attest to a strong relationship between the availability of energy, economic activity, and improvements in standards of living and overall social well-being [1].

Interest in satisfying the demand for energy arises from the basic goal of maintaining a certain level of human welfare and improving that welfare, wherever possible. Concern about this goal is indirectly expressed in the objective of obtaining the energy required to maintain or increase the growth of a nations' economy [2]. Knowing the required energy quantity provides better planning conditions for governments, especially for those dependent on outside energy sources (i.e. oil, natural gas, electricity). Moreover, the fast growth rate of the world's population and industrial development has resulted in environmental problems like an increase on green house gases and global warming.

Recent evidence indicates that road traffic emissions are a major source of air pollution with subsequent adverse human health effects [3, 4]. Although improvements in vehicle technology play a significant role in reducing traffic emissions at the source, air pollution abatement will remain a challenge because of increasing demand for transportation [5].

An increase in the overall vehicle-kilometers (veh-km) over the past two decades has outweighed any gains in emission reductions achieved through advances in car technologies [6, 7]. If planners are to implement policies that will reduce the need for

Z.W. Geem (Ed.): Music-Inspired Harmony Search Algorithm, SCI 191, pp. 163–172.
springerlink.com

travel and improve air quality caused by road traffic, then it is imperative not only to forecast the spatial pattern of traffic related to air pollution, but also to model energy consumption and transport demand. Modeling transport energy demand would enable calculating pollutants caused by transport.

Energy consumption in the transport sector usually depends on many factors such as vehicular usage, type of car, income, housing size, vehicular type and many other socioeconomic parameters. Including all the parameters in sectoral energy modeling is a difficult task since it requires much detailed study and also considerable data, for which many of the data are unavailable [8].

Robust estimations of transport energy issues may require a better selection of independent variables, such as socioeconomic indicators, to determine how much each parameter contributes to energy consumption. It also requires robust estimation techniques as well as better representations of independent variables. Therefore, estimations of transport energy demand based on the socio-economic and vehicle related parameters cannot be linear in nature. This is due to the general trend of the economic and vehicular factors. For example, it is better to represent the Gross Domestic Product (GDP) as a non-liner form. Thus, fitting non-linear models for estimating the energy demand may be better modeled.

The non-linearity of the economic indicators and veh-km to estimate energy demand lead to investigate the different form of solution approaches to a problem such as harmony search (HS). The available data are partly used for finding the optimal, or near optimal values of parameters, and for testing of the HS models. A robust model should be capable of displaying the sensitivity of energy consumption with the related indicators: population, GDP and veh-km. Meta-heuristic models are usually based on the assumption that a methodology is chosen and formed according to the minimization of some objective function. For this purpose, the HS algorithm is proposed to model energy consumption in the transport sector.

This chapter is organized as follows:

- Section 2: A literature survey of energy demand modeling is given.
- Section 3: The transport energy demand problem formulation is described.
- Section 4: Illustrative examples are given.
- Section 5: Conclusions are drawn.

2 Literature Survey of Energy Demand Modeling

For many countries, oil imports are usually the largest share of total imports since the transport sector is one of the major consumers of energy production in the world. Oil is the source of about one-fifth of the primary energy in the world [9]. It is also responsible for almost 60% of the energy consumption in Organization for Economic Co-operation and Development (OECD) countries and its rate of consumption is increasing in the developing world [10].

One of the main reasons for the increasing demand for energy use may be due to the rapid increase in population, mobility, businesses, globalization and transport demand. When a high level of energy demand and price levels is considered, oil represents one of the biggest shares of total energy consumption in many countries [11]. Studies on energy forecasting in Turkey are planned by the Ministry of Energy and

Natural Resources (MENR) and State Planning Organization (SPO) within five year development planning periods. The MENR model requires a large amount of detailed data for total and sectoral energy consumption estimations [12].

Meta-heuristic methods, which are used to solve combinatorial optimization problems, have been applied to estimate energy consumption in the energy literature. Genetic Algorithm (GA) based solution methods were used for estimating transport energy estimation [13-15]. Ozcelik and Hepbasli [16] applied a simulated annealing method to estimate energy production and consumption. The results of the study stated that a rising standard of living for the world's growing population meant increasing world energy consumption [17].

Lakshminarayana [18] described the pattern of electrical energy consumption on a medium/long term basis based on its interaction with other related techno-econometric and demographic indicators (variables), such as population, GDP, capital invested in the industrial sector, and gross oil and coal production. These indicators are taken as GDP, population, import and export for Turkey's energy demand estimation [19, 20].

Rijal et al. [21] developed an energy demand model which could project energy requirements in rural Nepal. The main end uses considered in this analysis were cooking, heating and lighting. Iniyan et al. [22] studied sensitivity analysis of an optimal renewable energy mathematical model on demand variations. Tukey's future energy demand estimation and optimal investment planning issues are studied by [23-27].

Gilland [28] developed an energy demand projection of the world for the years 2000 and 2020. Gungor and Arikan [29] developed a method to compare natural gas, imported coal and nuclear power plants in terms of long-term production economy.

Haldenbilen and Ceylan [11] developed three forms of energy demand equations in order to forecast the transport energy consumption for future projections based on GA approach. They used population, GDP and Car Equivalent Approach as independent variables. Ceylan et al. [8] proposed an HS based model for transport energy demand modeling problem. The results of the study showed that the HS algorithm may be used for energy modeling, but sensitivity analysis is required to obtain the best values of the HS parameters.

3 Meta-heuristic Harmony Search Algorithm

The HS algorithm, which has been developed by Geem et al. [30], mimics a musical improvisation process in which the musicians in an orchestra try to find a fantastic harmony through musical improvisations. This musical process can be adapted into engineering optimization processes where the main objective is to find the global or near-global solution of a given objective function. In this adaptation, each musician is replaced with a decision variable. Therefore, the perfect harmony means the global or near-global solution.

The HS algorithm has recently been applied to various engineering optimization problems including a river flood model [31], an optimal design of dam drainage pipes [32], a design of water distribution networks [33], a simultaneous determination of aquifer parameters and zone structures [34], a combined heat and power economic dispatch [35], an optimum design of steel frames [36], and a transport energy modeling problem [8]. HS was also applied to the optimal design of planar and space trusses [37-39].

The procedure of HS is composed of five steps:

Step 1: Initialization of the problem and algorithm parameters

$$\text{Minimize } f(\text{x}), \quad \text{subject to } x_i \in \mathbf{X}_i, \; i = 1, 2, \ldots, N \tag{1}$$

where $f(x)$ is the objective function to be minimized, x is the set (so called orchestra) of decision variables x_i; N is the number of decision variables (music instruments); and $\mathbf{X_i}$ is the set of the possible range of values for each decision variable (the pitch range of each instrument).

Step 2: Initialization of the harmony memory

The harmony memory (HM) is a memory location where all the solution vectors and corresponding objective function values are stored. In this step, the HM matrix is filled with solution vectors by taking the possible ranges into account and fitness function values are calculated for each solution vector.

Step 3: Improvisation of a new harmony

A new harmony vector, $\mathbf{x}' = (x'_1, x'_2, \ldots, x'_N)$, is generated based on three rules. They are memory consideration, pitch adjustment, and random selection. The value of a design variable can be selected from the values stored in HM with a probability of harmony memory considering rate (HMCR). It can be further adjusted by moving to a neighbor value of a selected value from the HM with a probability of pitch adjusting rate (PAR). Or, it can be selected randomly from the set of all candidate values without considering the stored values in HM, with the probability of (1-HMCR). The improvisation process of the HS algorithm is depicted in Figure 1.

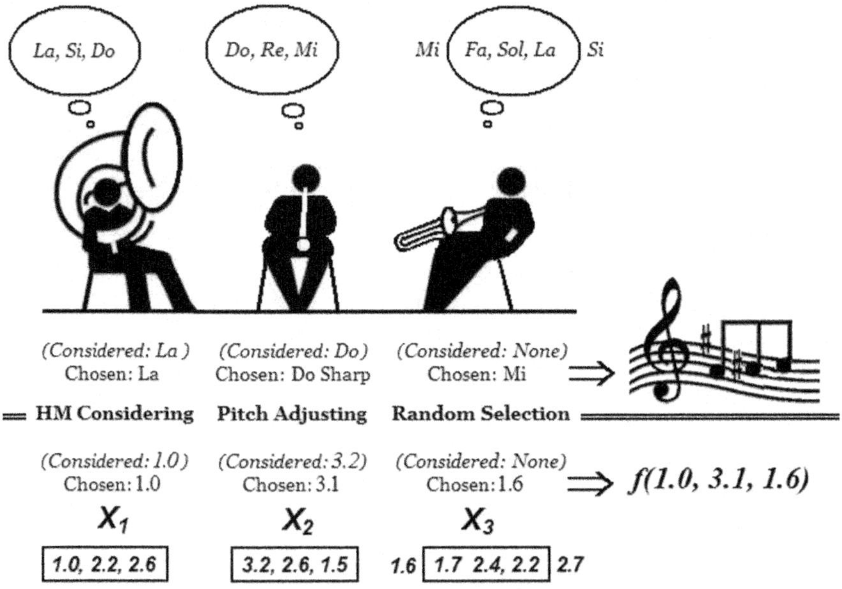

Fig. 1. Improvisation of a new harmony in HS

As can be seen in Figure 1, each musician in the orchestra has some notes in their memories. They intend to improvise a new harmony with the question of "which note will be played?" in their minds. Three rules of improvising a new harmony can be explained using the following way:

i) Memory Consideration
The first musician in the figure has a memory containing 3 notes, {La, Si, Do}. He decides to choose *La* in his memory and plays it, directly. Likewise, if the first decision variable represents the first musician, the value of 1.0 can be chosen from a memory.

ii) Pitch Adjustment
The second musician also has a 3-notes memory, {Do, Re, Mi}. Differently from the first musician, he first chooses the note *Do*. Then, he can play its neighbor pitch such as *Do Sharp*. Likewise, the value of 3.2 is chosen from the memory, and then it can be adjusted into a neighbor value 3.1 in this step.

iii) Random Selection
The last musician has some notes in his memory, {Fa, Sol, La}, as well. Although this memory was used during the past improvisations, due to his musical knowledge he can also play all possible pitches, {Do, Re, Mi, Fa, Sol, La, Si, Do+}. Thus, when he decides to play a note randomly, he can choose any of these notes, *Mi* in this example. As being in the possible data set, 1.6 can be chosen in this step, randomly, even if it doesn't exist in the memory.

After each musician has decided what to play, the new harmony is composed of (La, Do#, Mi). Similarly, a new solution vector is determined as (1.0, 3.1, 1.6).

Step 4: Update the HM
If the newly generated harmony vector gives a better function value than the worst one, the new harmony vector is included in the HM and the worst harmony is excluded.

Step 5: Check the termination criterion
If the stopping criterion (i.e. maximum number of improvisations) is satisfied, computation is terminated. Otherwise, Steps 3 to 5 are repeated.

4 Transport Energy Demand Problem

Due to its big share, transport energy consumption is an important component of total energy consumption for countries. Modeling this component plays vital importance for future planning studies. Transport energy planning is not possible without a reasonable knowledge of past and present energy consumption and likely future demands [11].

The estimation of future energy demand can be represented with various forms of mathematical expressions. The non-linear forms of the equations may better estimate the future energy demand of Turkey due to the fluctuations of socio-economic indicators. For modeling the energy demand in transport sector, Harmony Search Transport

Energy Demand Estimation (HASTEDE) models have been proposed [8] and they are in the following structures:

$$f(x)_{linear} = w_1 X_1 + w_2 X_2 + w_3 X_3 + w_4 \tag{2}$$

$$f(x)_{exp} = w_1 X_1^{w_2} + w_3 X_2^{w_4} + w_5 X_3^{w_6} + w_7 \tag{3}$$

$$f(x)_{quad} = w_1 X_1 + w_2 X_2 + w_3 X_3 + w_4 X_1 X_2 + w_5 X_1 X_3 + w_6 X_2 X_3 + w_7 \tag{4}$$

where X_1, X_2 and X_3 are the GDP (10^9\$), population (10^6), and total annual veh-km (10^9), respectively; $w_i \in \mathbf{W}_i \; (i = 1,2,3,...,N)$ are the corresponding weighting factors. To minimize the sum of squared error (SSE) between the observed and estimated values, the optimization function Z is defined as:

$$\text{Minimize} \quad Z = \sum_{1}^{m} \left(TED_{actual} - TED_{predicted} \right)^2 \tag{5}$$

where TED_{actual} and $TED_{predicted}$ are the actual and predicted transport energy demand, and m is the number of observations.

4.1 Data for HASTEDE Models

The GDP and the sectoral energy consumption (SEC) data in the following figure were collected from the Central Bank of Turkey [40] and the WEC-TNC [12]. Observed veh-km was taken from the GDTH [41]. The observed general trend of energy demand GDP, population, and total veh-km between 1970 and 2005 can be seen in Figure 2.

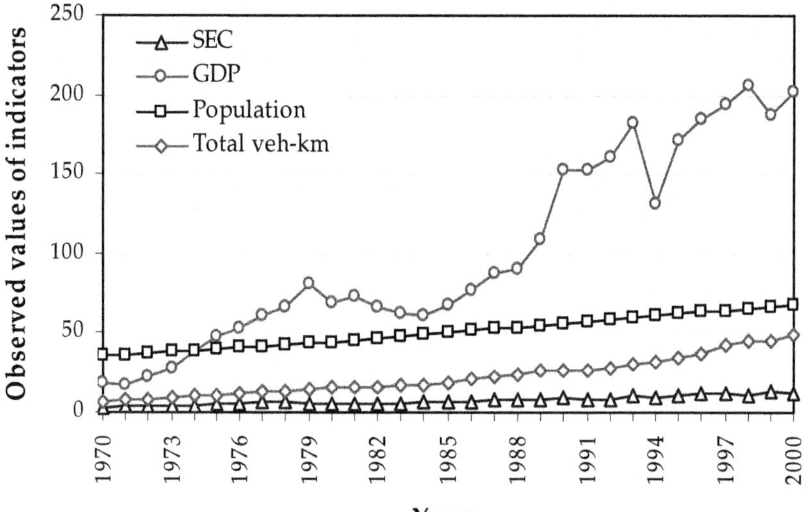

Fig. 2. General trend of SEC and related parameters

4.2 HASTEDE Applications

For the application of each HASTEDE model for estimating transport energy demand, the HS parameters are set as in Table 1. Lee et al. [39] have recommended the HS parameter values: HMCR range between 0.7 and 0.95; PAR values range between 0.2 and 0.5; and HMS values range between 10 and 50 to produce good performance of the HS algorithm. Thus, they are taken within the bounds of recommended values.

Table 1. HS parameters for HASTEDE models

	HASTEDE_{lin}	HASTEDE_{exp}	HASTEDE_{quad}
HMS		20	
HMCR		0.90	
PAR		0.40	
Number of weighting variables	4	7	7
Number of improvisations		100,000	

The solution of the linear form of the HASTEDE model is:

$$\text{HASTEDE}_{lin} = 0.0077X_1 - 0.0351X_2 + 0.2473X_3 + 2.999 \tag{6}$$

The solution of the exponential form of the HASTEDE model is:

$$\text{HASTEDE}_{exp} = 1.7727X_1^{0.1822} - 0.3796X_2^{0.7974} + 0.4525X_3^{0.9297} + 4.2140 \tag{7}$$

The solution of the quadratic form of the HASTEDE model is:

$$\begin{aligned}\text{HASTEDE}_{quad} = {} & 0.0888X_1 - 0.9775X_2 - 0.0718X_3 - 0.0019X_1X_2 \\ & + 0.0008X_1X_3 + 0.0062X_2X_3 + 4.8984\end{aligned} \tag{8}$$

Solution of the above models is performed for 20 solution vectors and it is performed according to the procedure defined Section in 3 for the period of 1970-1995 in order to obtain weighting parameters for each of the HASTEDE models. Convergence of models is given in Figure 3 in logarithmic scale. As can be seen in the figure, the HASTEDE models showed a steady convergence for this problem.

About two-thirds of the data are used for estimating weighting parameters and one-third of the data are used for testing the HASTEDE models. The relative errors obtained for the example in 2001 is about 12, 16, 31, 45 percent for linear, exponential, quadratic, and MENR, respectively, but relative errors in 2005 are respectively 14, 12, 17, 42 percent.

The values of the objective function (Eq. 5) are given in Table 2. The quadratic form of the HASTEDE model provides the lowest SSE, and it may be used for future energy prediction for planning purposes.

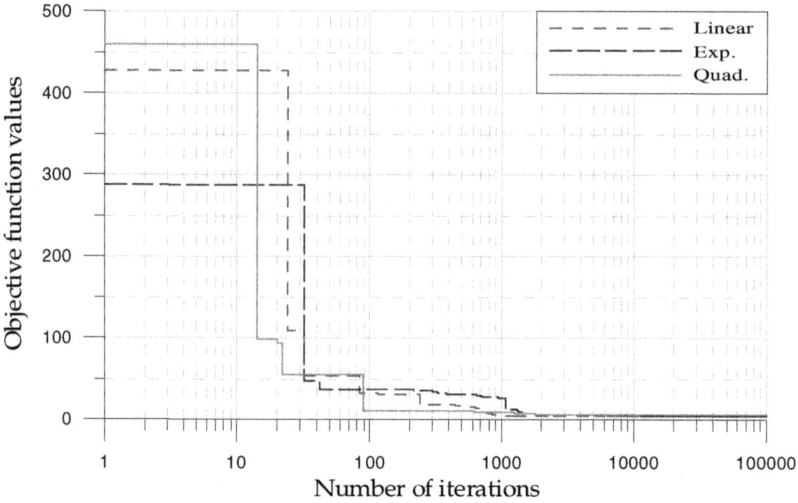

Fig. 3. Convergence history of the HASTEDE models

Table 2. HASTEDE model results

Type	$SSE_{HASTEDE}$
Linear	6.46
Exponential	5.94
Quadratic	4.34

5 Conclusions

This chapter reviewed the application of the HS optimization algorithm to the solution of the transport energy demand modeling problem. Formulation of the HS method and its application to this problem were provided. Three forms of the forecasting models were developed and successfully solved with the HS procedure.

References

1. Nathwani, J.S., Siddall, E., Lind, N.C.: Energy for 300 years. Institute for Risk Research, University of Waterloo, Waterloo (1992)
2. Dincer, I., Dost, S.: Energy and GDP. Int. J. Energy Res. 21, 153–167 (1997)
3. Faiz, A.: Automotive emissions in developing countries-relative implications for global warming, acidification and urban air quality. Transportation Research (Part A) 27, 167–186 (1993)
4. Colvile, R.N., Hutchinson, E.J., Mindell, J.S., Warren, R.F.: The transport sector as a source of air pollution. Atmospheric Environment 35, 1537–1565 (2001)
5. World Business Council for Sustainable Development, World mobility at the end of the twentieth century and its sustainability. WBCSD, Geneva (2001)

6. Anderson, W.P., Kanaroglou, P.S., Miller, E.J.: Urban form, energy and the environment: A review of issues, evidence and policy. Urban Studies 33, 7–35 (1996)
7. Potoglou, D., Kanaroglou, P.S.: Carbon monoxide emissions from passenger vehicles: predictive mapping with an application to Hamilton, Canada. Transportation Research (Part D) 10, 97–109 (2005)
8. Ceylan, H., Ceylan, H., Haldenbilen, S., Baskan, O.: Transport energy modeling with meta-heuristic harmony search algorithm, an application to Turkey. Energy Policy 36, 2527–2535 (2008)
9. Asmann, D., Sieber, N.: Transport in developing countries: renewable energy versus energy reduction. Transport Reviews 25, 719–738 (2005)
10. International Energy Agency (IEA), World energy outlook 2005: middle east and north africa insights. OECD/IEA Publishers, Paris (2005)
11. Haldenbilen, S., Ceylan, H.: Genetic algorithm approach to estimate transport energy demand in Turkey. Energy Policy 33, 89–98 (2005)
12. World Energy Council-Turkish National Committee, Energy Statistics. WEC-TNC, Ankara (2006) (in Turkish)
13. Ceylan, H., Ozturk, H.K.: Estimating energy demand of Turkey based on economic indicators using genetic algorithm approach. Energy Conversion and Management 45, 2525–2537 (2004)
14. Canyurt, O.E., Ozturk, H.K.: Three different applications of genetic algorithm search techniques on oil demand estimation. Energy Conversation and Management 47, 2148–3138 (2006)
15. Canyurt, O.E., Ozturk, H.K., Hepbasli, A., Utlu, Z.: Genetic algorithm approaches for the transport energy demand estimation: model development and application. Energy Sources (Part A) 28, 1405–1413 (2006)
16. Ozcelik, Y., Hepbasli, A.: Estimating petroleum energy production and consumption using a simulated annealing approach. Energy Sources (Part B: Economics, Planning and Policy) 1, 255–265 (2006)
17. Gilland, B.: Energy for the 21st century: an engineer's view. Endeavour 14, 80–86 (1990)
18. Lakshminarayana, G.S.: Importance of long range energy demand forecasting in power system planning. Journal of Irrigation and Power 48, 29–38 (1991)
19. Unler, A.: Improvement of energy demand forecasts using swarm intelligence: The case of Turkey with projections to 2025. Energy Policy 36, 1937–1944 (2008)
20. Toksari, M.D.: Ant colony optimization approach to estimate energy demand in Turkey. Energy Policy 35, 3984–3990 (2007)
21. Rijal, K., Bansal, N.K., Grover, P.D.: Rural household energy demand modeling. Energy Economics 10, 279–288 (1990)
22. Iniyan, S., Sumathy, K., Suganthy, L., Samuel, A.A.: Sensitivity analysis of optimal renewable energy mathematical model on demand variations. Energy Conversation & Management 41, 199–211 (2000)
23. Demirbas, A.: Energy balance, energy sources, energy policy, future developments and energy investments in Turkey. Energy Conversion and Management 42, 1239–1258 (2001)
24. Yumurtaci, Z., Asmaz, E.: Electric energy demand of Turkey for the year 2050. Energy Sources 26, 1157–1164 (2004)
25. Sozen, A., Arcaklioglu, E., Ozkaymak, M.: Modeling of the Turkey's net energy consumption using artificial neural network. International Journal of Computer Applications in Technology 22, 130–136 (2005)
26. Akay, D., Atak, M.: Grey prediction with rolling mechanism for electricity demand forecasting of Turkey. Energy 32, 1670–1675 (2007)

27. Sozen, A., Arcaklioglu, E.: Prospects for future projections of the basic energy sources in Turkey. Energy Sources (Part B: Economics, Planning, and Policy) 2, 183–201 (2007)
28. Gilland, B.: Population, economic growth, and energy demand, 1985–2020. Population and Development Review 14, 233–244 (1988)
29. Gungor, Z., Arikan, F.: A fuzzy outranking method in energy policy planning. Fuzzy Sets and Systems 114, 115–122 (2000)
30. Geem, Z.W., Kim, J.H., Loganathan, G.V.: A new heuristic optimization algorithm: harmony search. Simulation 76, 60–68 (2001)
31. Kim, J.H., Geem, Z.W., Kim, E.S.: Parameter estimation of the nonlinear Muskingum model using harmony search. Journal of the American Water Resources Association 37, 1131–1138 (2001)
32. Paik, K., Kim, J.H., Kim, H.S., Lee, D.R.: A conceptual rainfall-runoff model considering seasonal variation. Hydrological Processes 19, 3837–3850 (2005)
33. Geem, Z.W.: Optimal cost design of water distribution networks using harmony search. Engineering Optimization 38, 259–280 (2006)
34. Ayvaz, T.: Simultaneous determination of aquifer parameters and zone structures with fuzzy c-means clustering and meta-heuristic harmony search algorithm. Advances in Water Resources 30, 2326–2338 (2007)
35. Vasebi, A., Fesanghary, M., Bathaee, S.M.T.: Combined heat and power economic dispatch by harmony search algorithm. International Journal of Electrical Power & Energy Systems 29, 713–719 (2007)
36. Degertekin, S.O.: Optimum design of steel frames using harmony search algorithm. Struct. Multidisc. Optim. 36, 393–401 (2008)
37. Lee, K.S., Geem, Z.W.: A new structural optimization method based on the harmony search algorithm. Computers and Structures 82, 781–798 (2004)
38. Lee, K.S., Geem, Z.W.: A new meta-heuristic algorithm for continuous engineering optimization: harmony search theory and practice. Computational Methods in Applied Mechanical Engineering 194, 3902–3933 (2005)
39. Lee, K.S., Geem, Z.W., Lee, S.H., Bae, K.W.: The harmony search heuristic algorithm for discrete structural optimization. Engineering Optimization 37, 663–684 (2005)
40. Central Bank of Turkey (CBT) (2007) (accessed December 11, 2008) (in Turkish), `http://www.tcmb.gov.tr`
41. General Directorate of Turkish Highways (GDTH), Annual statistics of Turkish highways (2005) (in Turkish) (accessed December 11, 2008), `http://www.kgm.gov.tr`

Sound Classification in Hearing Aids by the Harmony Search Algorithm

Enrique Alexandre, Lucas Cuadra, and Roberto Gil-Pita

Department of Signal Theory and Communications, University of Alcalá, 28805-Alcalá de Henares, Spain
{enrique.alexandre,lucas.cuadra,roberto.gil}@uah.es

Abstract. This chapter focuses on the application of the harmony search algorithms to the problem of selecting more appropriate features for sound classification in digital hearing aids. Implementing sound classification algorithms embedded in hearing aids is a very challenging task. Hearing aids have to work at very low clock frequency in order to minimize power consumption, and thus maximize battery life. This necessitates the reduction of computational load while maintaining a low error probability. Since the feature extraction process is one of the most time-consuming tasks, selecting a reduced number of appropriate features is essential, thus requiring low computational cost without degrading the operation. The music-inspired harmony-search (HS) algorithm allows for effectively searching adequate solutions to this strongly constrained problem. By starting with an initial set of 74 different sound-describing features, a number of experiments were carried out to test the performance of the proposed method. Results of the harmony search algorithm are compared to those reached by other widely used methods.

Keywords: Sound Classification, Feature Selection, Hearing Aids.

1 Introduction

Hearing aids are electronic instruments that aim at compensating the acoustic loss from which hearing impaired people suffer. Yet only 20% of hearing-impaired people who could benefit from digital hearing aids actually purchase one, and of these, about 25% do not wear the devices [1]! This is believed to be due, at least in part, to the unpleasant amplification of background noise encountered in everyday life. The speech signal is the primary target and concern for most of users. However, there are many other different sound sources such as traffic, school, people shouting in a football stadium, etc. Therefore, in order to successfully face the variety of acoustic environments represented in Figure 1, most hearing aid users generally prefer to have a number of amplification schemes [2, 3].

A particular application that would be considered as very useful for hearing aid users, especially by the elderly, is that in which the hearing aid itself classifies the acoustic environment that surrounds the user, and automatically selects the amplification 'program' that is best adapted to such surroundings. This is referred to as self-adaptation. Some digital hearing aids allow the user to manually select the program adapted to the acoustic environment he/she is in. This approach commonly exceeds the abilities of most hearing aid users, especially the elderly. In a second feasible line of attack, an automatic sound classifier can be embedded in the hearing aid so that it assists it in selecting the proper program.

Z.W. Geem (Ed.): Music-Inspired Harmony Search Algorithm, SCI 191, pp. 173–188.
springerlink.com

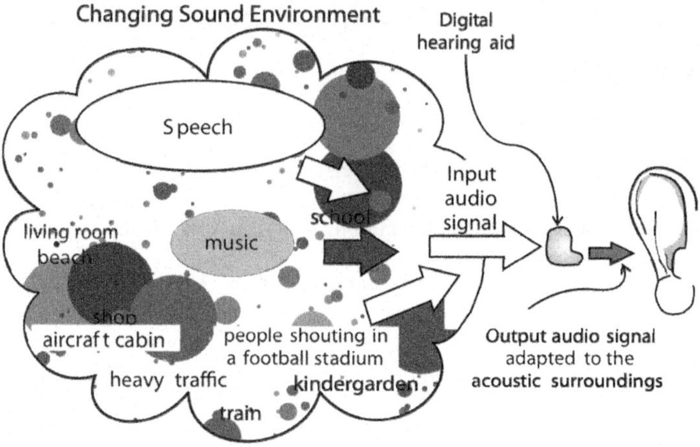

Fig. 1. The variety of sounds in everyday life

An automatic sound classifier would potentially be a very useful application. However, at present, many of the medium-cost hearing aids in the market cannot automatically adapt to the changing acoustical environment. In spite of the impressive advances in microelectronics, digital hearing aids suffer yet from severe design trade-offs, which, in the effort of making this chapter stand by itself, are briefly described in Section 0. These limitations mainly arise from the need of reducing the power consumption in order to maximize the battery life. As a consequence, implementing sound classification algorithms embedded in hearing aids becomes a very challenging task [4-14]. The underlying reason is that it requires optimizing each parameter for the sound classifier in order to reduce its computational complexity while maintaining a low error probability, and thus, a good dynamic adaptation to the sound surroundings.

In particular, the feature extraction process, essential to properly characterize the sound to be classified, is one of the most time-consuming tasks. Therefore, reducing the number of features is probably one of the most successful ways of diminishing the complexity of the classification system. The problem here is that, as will be commented in Section 3, an excessively small number of features may increase error probability. Thus, selecting those more appropriate features [15] is a crucial point for the application at hand. Here, 'appropriate' means within this framework that they should have low computational cost (i.e. a reduced number of mathematical operations per second to be computed) without degrading the hearing aid operation. Just in this effort, this chapter explores the feasibility of using the harmony search (HS) algorithm [16-21] to select the more satisfactory features constrained to the severe limitations previously mentioned.

As it was commented above, hearing loss is currently an important health problem. Approximately 13% of the population in highly developed countries suffers from considerable hearing loss whose negative consequences could be mitigated by using some kind of hearing aid [1-14]. These hearing losses strongly affect the speech communication and disqualify many hearing-impaired people from normal life activities. Hearing impairment is more prevalent with increasing age. For instance, almost 50% of those elder over the age of 75 are 'hard of hearing'. Taking into account that the population of most highly developed countries is quickly aging, and that hearing impairment has an important impact on communication, society faces a serious health and social problem. This compels scientists and engineers to develop more advanced digital hearing aids, and, most importantly, to make them comfortable and easy to handle, especially for elderly people.

Despite the fact that modern digital hearing aids exhibit the potential to be designed and programmed to compensate for most specific hearing losses, a barrier to achieving this aim is the user's comfort and subjective preferences. Only 20% of hearing impaired people who could benefit from a hearing aid acquires the device. And, as mentioned before, approximately 25% of them do not wear their hearing aids.

What is the explanation underlying these facts? Apart from the high costs of the most advanced hearing aids (which may provide higher fidelity of sound, greater overall amplification, directional sound detection, dynamic compression and frequency-specific amplification), there is still a critical, scientific and technological drawback that has not yet been solved effectively and comfortably at medium-low cost: the automatic adaptation to the changing acoustic environment.

As shown in Figure 1, this problem arises from the fact that hearing aid users face a variety of different acoustic environments in their daily life, e.g. quiet conversation, speech in a noisy cafe, traffic, loud music, etc.

Unfortunately, most of medium-low cost hearing aids are usually designed for only one listening environment. The overall performance to the individual is unsatisfactory. The unit is either too loud or too soft or the quality is poor. For example, entering a crowded café produces a sudden, uncomfortable, irritating, and amplified noise. Many times the patient can 'hear' but not understand the speech signal. With respect to the diversity of sound environments the patient has to face on in his/her normal life, it has been shown that hearing aid users normally prefer to have a number of amplification schemes fitted to different listening conditions [1-3]. To fulfill such a need there are two essential approaches in digital hearing aids. The first approach allows the customer to manually select –among a variety of programs (different frequency responses or other processing options such as compression methods, directional microphone, feedback canceller, etc.)– the one the customer considers more adequate to the acoustic surroundings. The user has to be familiar with the acoustic environment and determine which program best fits the circumstance by using a switch on the hearing instrument or some kind of remote control. Nonetheless this approach exceeds the technical capabilities of most hearing aid users, in particular for the smallest 'in-the-canal' (ITC) or 'completely-in-the-canal' (CIC) hearing aids, which are virtually hidden to the casual viewer.

The second approach, which could appreciably get better usability of the hearing aid, is that in which the hearing device itself selects the most suitable program. In this paper, an HS algorithm is used to select the sound describing features

2 Framework: Design Constraints

As mentioned in the introduction, digital hearing aids present substantial design constraints. Figure 2 displays the typical and critical elements of the structure of a digital hearing aid. The basic components are:

- A microphone, to convert the input sound into an electric signal
- A number of electronic blocks aiming at compensating the hearing loss. Among other signal processing functionalities, a proper sound amplifier will intensify the strength of the signal in which the user is interested (e.g. speech, in most of the cases)
- A tiny loudspeaker to convert the electrical signal back to sound
- A coupler to the ear canal, usually referred to as an 'ear mold', and
- A battery to provide electric power to the electronic devices of the hearing instrument.

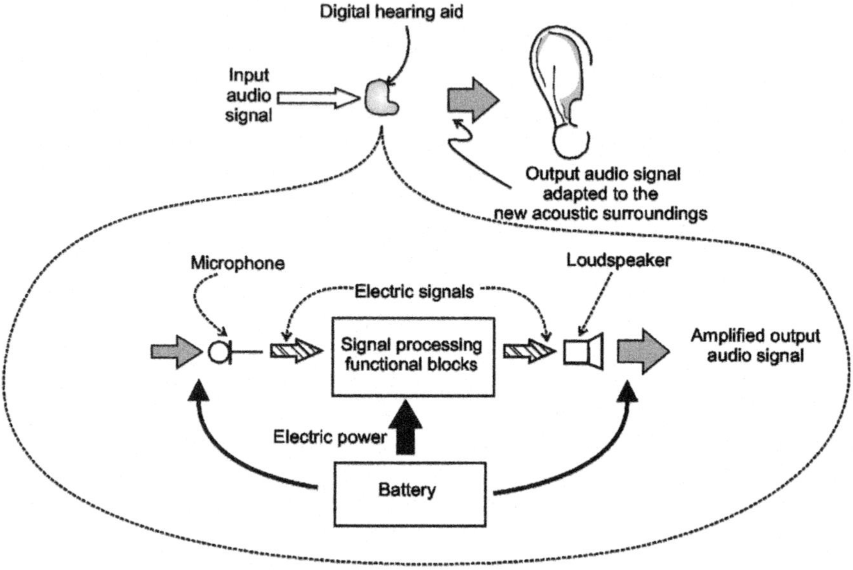

Fig. 2. Typical structure of a digital hearing aid

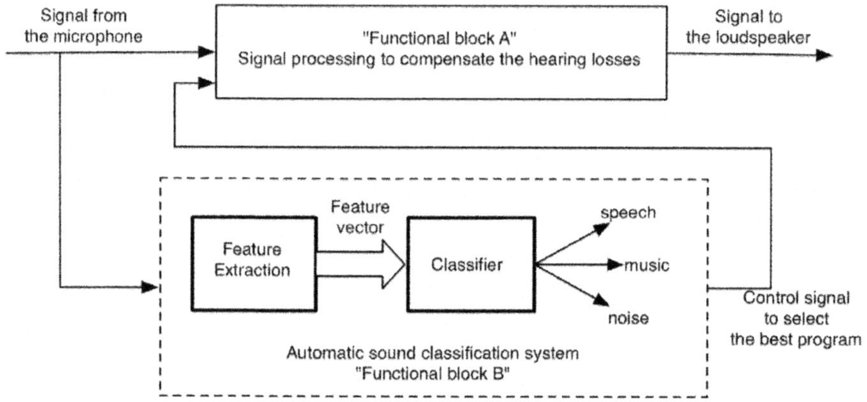

Fig. 3. Representation of the functional blocks to be implemented on the digital hearing aid

The signal processing electronic stages represented in Figure 2 are based on a Digital Signal Processing (DSP) integrated circuit. The DSP performs a number of arithmetical operations. Figure 3 illustrates the two main functional blocks that must be implemented on the DSP. The first one, labeled 'functional block A', corresponds to the set of signal processing stages aiming to compensate the hearing losses. The second one, 'functional block B', is the classification system itself. Note that the aforementioned classifying system consists conceptually of two basic parts:

- A feature extraction stage in order to properly characterize the signal to be classified. (See Sections 3 and 5.2).
- The three-class (speech, music and noise) classifier (See Section 5.3).

The aforementioned restrictions arise mainly from the tiny size of the hearing aid (especially for the ITC and CIC models), which, as illustrated in Figure 2, must contain not only the electronic equipment but also the battery. In fact the DSP usually integrates the A/D and D/A converters, the filter bank, RAM, ROM and EPROM memory, input/output ports, and the core. The immediate consequence is that the hearing aid has to work at very low clock frequencies in order to minimize the power consumption and thus maximize the life of its battery. The power consumption must be low enough to ensure that neither the size of the battery pack nor the frequency of battery changes will annoy the user.

Another key point to note regarding this is that a considerable part of the computational capabilities are already used for running the algorithms of 'functional group A' aiming to compensate the acoustic loss. Therefore, designers are constrained to use the remaining part to implement the embedded sound classifier. Roughly speaking, the computational power available does not exceed 3 Million Instructions Per Second (MIPS), with only 32 Kbytes of internal memory. The time/frequency decomposition is performed by using an integrated Weighted Overlap-Add (WOLA) filter bank, with 64 frequency bands. It must be considered that only the filter bank requires about 33% of the computation time of the DSP. Since functional block A requires another 33%, the immediate consequence is that only about 1 MIPS is free for designing the classifier. Within the framework imposed by this constraint, the success of the classification strongly depends upon the sound describing features used to characterize the signal to be classified.

3 Fundamentals of Selection and Extraction of Features

3.1 Intuitive Approach

As previously stated, these processes aim at finding and extracting the kind of information of the audio signal that should assist the classifier in distinguishing between the acoustical classes. The performance of the system is severely dependent on what features are used. For instance, the classifier may not perform properly if its input features do not contain all this essential, adequate information. Additionally, the features not only have to properly describe the sound (allowing its subsequent classification), but also have to make efficient use of the DSP resources. This means that computing

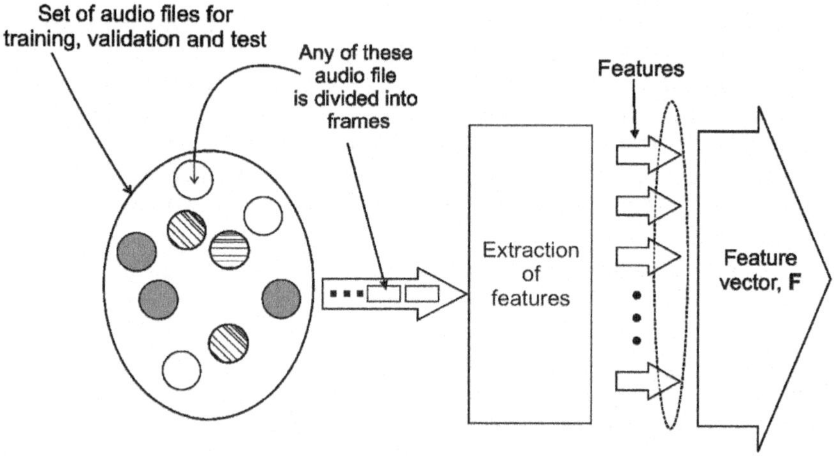

Fig. 4. The feature extraction process

the features on the DSP core should require a sufficiently small number of basic mathematical operations per second. The process of computing the features is called 'feature extraction', and is displayed in Figure 4.

The feature extraction requires dividing the sound signal into shorter parts called 'frames'. After a sequence of mathematical operations, the feature extractor produces a number of sound describing features that form the vector **F**, the information of which the classifier makes use.

The feature extraction process can be the most time-consuming task in the classification system. It is very important therefore to carefully design and select the features that will be calculated, taking into account the limited facilities available in the DSP. An illustrative example of the consequences of these constraints can be found in the fact that the strong computational limits restrict the frequency resolution of the time/frequency transforms which are necessary for those that make use of 'spectral information'. Thus, the selection of an appropriate number of features becomes one of the key topics involved in the successfully design of the classifier. This trade-off consists of reducing the number of features while maintaining the probability of correct classification, and the speech intelligibility perceived by the user. As illustrated in Figure 4, this is equivalent to diminishing the dimension of a vector **F** composed by the selected features.

Classifying the audio signal entering the hearing aid as belonging to speech, music or noise, requires the unit to extract (compute) features containing the relevant information that should assist the classifier in distinguishing between the classes. For illustrative purposes, we have grouped these two tasks into the so-called Functional Group B illustrated in Figure 3. It is necessary for classifying the input sound to carry out a number of signal processing steps summarized in subsection 3.2, and illustrated in Figure 4.

3.2 Mathematical Characterization of the General Feature Extraction Process

The input audio signal to be classified, $X(t)$, which will be assumed as a stochastic process, is segmented into frames as follows:

$$X_i(t), \quad i = 1, \cdots, r \tag{1}$$

where r is the number of frames into which the signal has been divided.

The next step consists in sampling any frame $X_i(t)$:

$$X_i(t_k), \quad k = 1, \cdots, p \tag{2}$$

where p is the number of samples per frame.

Since each frame $X_i(t)$ is a windowed stochastic process, any of its samples, $X_i(t_k)$, is a random variable, labeled $X_{i\,k}$.

Thus, for each audio frame, $X_i(t)$, the following random vector is obtained:

$$\mathbf{X_i} = \lfloor X_{i1}, \cdots, X_{ip} \rfloor \tag{3}$$

This is just the initial information the system uses to calculate all the features describing frame $X_i(t)$.

With respect to the aforementioned features, let

$$\Gamma = \{ f_1, \cdots, f_{N_F} \} \tag{4}$$

be the set that contains all the available features, N_F being the number of features.

Any feature f_k can be assumed, in the most general case, as a complex function of p complex variables:

$$f_k : C^p \to C \tag{5}$$

Since frame $X_i(t)$ has been shown to be the random-variable vector in Eq. 3, then any feature f_k applied on it, $f_k(\mathbf{X_i})$, is thus a function of p random variables, $f_k(\mathbf{X_{i1}}, ..., \mathbf{X_{ip}})$, and, consequently, a random variable. In order to simplify the notation, the random variable, $f_k(\mathbf{X_{i1}}, ..., \mathbf{X_{ip}})$, will be labeled f_{ki}.

Finally, for completing the characterization of the input audio signal, the aforementioned sequence of processes has to be applied onto all the r frames into which the input audio signal has been segmented.

The next step assesses how feature $f_k \in \Gamma$ describes the input signal. One of the results that provides the previously described sequence of processes is the random data vector

$$[f_{k1}, \cdots, f_{kr}] \equiv \mathbf{F_k} \tag{6}$$

The elements of this vector are the results obtained when feature f_k is applied to each of the r frames into which the input audio signal has been segmented to be processed. The random vector $\mathbf{F_k}$ can be characterized for example by estimating its mean value:

$$\hat{E}[\mathbf{F_k}] \tag{7}$$

and its variance:

$$\hat{\sigma}^2[\mathbf{F_k}] \tag{8}$$

In Eqs. 7 and 8, it has been used, respectively, the mean and variance statistical operators

$$\hat{E}[Y] \equiv \mu \equiv \frac{1}{M}\sum_{m=1}^{M} y_m \tag{9}$$

$$Var[Y] \equiv \hat{\sigma}^2 \equiv \frac{1}{M}\sum_{m=1}^{M} (y_m - \mu)^2 \tag{10}$$

where M represents the number of values of a given random variable Y.

Note that the previously described statistical characterization can be done for all the available features $f_k \in \Gamma$ so that the feature extraction algorithm generates

$$\mathbf{F} = \left[\hat{E}[\mathbf{F}_1], \left[\hat{\sigma}^2[\mathbf{F}_1]\right], \cdots, \hat{E}[\mathbf{F}_{N_F}], \left[\hat{\sigma}^2[\mathbf{F}_{N_F}]\right] \right] \tag{11}$$

a random vector whose dimension is $dim(\mathbf{F}) = 2N_F = m$.

This is just the signal-describing vector that should feed the classifier. For the sake of clarity, it is written as $\mathbf{F} = \left[F_1, \cdots, F_{N_F}\right]$.

Finally, taking this data structure in mind, the immediate goal is:

Given the set of N_F available, general-purpose features,

$$\Gamma = \{f_1, \cdots, f_{N_F}\} \tag{12}$$

to select a subset $\Phi \subseteq \Gamma$ that minimizes the mean error probability of correct classification.

Finding the solution requires, for any candidate feature subset $\Phi \subseteq \Gamma$, the following sequence of operations:

- Feeding the complete system with input audio files from a proper database, which will be described in Section 5.1
- Calculating the feature vector for the complete file, as previously explained
- Classifying, and
- Calculating the mean squared error (MSE) between the classifier output and the correct output, and the probability of correct classification.

As shown, the process of selecting a subset containing those more appropriate features, those that minimize the classification error probability, is very complex. The feasibly of the HS method is now explored.

4 Harmony Search Algorithm in a Nutshell

Harmony search (HS) algorithm is a meta-heuristic searching procedure inspired by the processes of music creation and improvisation. In these activities, musicians look for the 'best' combination of sounds in order to produce esthetically pleasant compositions [16-21]. In such improvisation process of composition, the musician remembers those pieces of music that sound melodiously and that can be used in order to inspire novel compositions. Thus, in the effort of exploring utterly different melodies, the musician is able to generate 'creative' compositions.

The HS algorithm is an optimization method that mimics the composition rules in order to search for the best solution to a mathematical problem by minimizing some kind of 'fitness function'. In this music-inspired framework, each solution is considered a 'melody', and its fitness represents its musical quality. The musician creates new combinations of sounds, and memorizes the best M melodies, i.e. those with the highest quality. When the musician has to improvise a novel composition, he/she either creates an entirely novel melody with a given probability, let's say, for instance, P_n, or selects one of the previous memorized melodies with probability $1-P_n$. Please note that, in the effort of producing new variations, any piece of music is in fact randomly modified in some of its parts. The iterative application of this process allows searching for those compositions that minimize the fitness function, or in other words, those that 'sound better' [16-21].

The question now is: What is the analogy to the problem of selecting those more appropriate sound-describing features for the application at hand? In the music-inspired framework previously mentioned, the key point consists in defining the basic structure of a melody. In the problem of selecting features for sound classification in hearing aids, the set of features to be selected must be coded in an appropriate way. In this work they have been encoded by using a bitstream so that each bit determines if a given feature is used or not: bit '1' means that the considered feature will be selected, while bit '0' represents the contrary. As a consequence, and if there is N_F available features, each melody/composition will consist of a set of N_F binary values. With this in mind, the way the HS algorithm works for the problem at hand can be described as follows:

Step 1. Preprocessing stage: The harmony memory is filled with random compositions and the associated fitness values are evaluated.

Step 2. A new melody is created. Any melody can be generated either at random (this case is selected with a probability P_n), or by copying it from one memorized melody (this alternative case is selected with a probability equal to $1-P_n$).

Step 3. The components of the novel melody are then modified with a probability of P_m.

Step 4. The new melody is then evaluated and scored, taking into account the fitness function defined in order to carry out the selection process. In this chapter, the validation error rate has been selected as fitness function.

Step 5. If the novel melody scores better than the worse one included in the memory, then it is stored, replacing the worse one. This process iterates beginning in Step 2.

Please note that using the HS optimization algorithm may potentially provide a good combination of features. The great advantage is that it prevents the process from evaluating all the possible 2^{N_F} combinations of N_F features, which, in some cases, could demand the use of excessive computational resources.

5 Experimental Setup

Prior to the description of the experiments carried out and the discussion of the corresponding results, we now describe the database used (Section 5.1), the particular feature extraction process, the kind of classifier, and the parameters of the search algorithms used for illustrating this chapter.

5.1 Available Database

The sound database used for the experiments consisted of a total of 2,627 files, with a length of 2.5 seconds each. The sampling frequency was 22,050 Hz with 16 bits per sample. The files correspond to the following categories: speech, music and noise. Noise sources were varied, including those corresponding to the following environments: aircraft, bus, cafe, car, kindergarten, living room, nature, school, shop, sports, traffic, train, and train station. Music files were both vocal and instrumental. The files with speech in noise presented different Signal to Noise Ratios (SNRs) ranging from 0 to 10 dB.

The database has been divided into three different sets for training, validation and test, including 943 (35%), 405 (15%) and 1,279 (50%) files respectively. The division has been made randomly, ensuring that the relative proportion of files of each category is preserved for each set.

5.2 Feature Extraction Stage

As described in Section 3.2, the particular feature extraction carried out in the experiments may be summarized as follows:

1. The input signal is divided into frames with a length of 512 samples (23.22 ms for the considered sampling frequency) with no overlap between adjacent frames.
2. The Discrete Cosine Transform (DCT) is computed [14].
3. All considered features are calculated.
4. Finally, the mean and standard deviation values are computed every 2.5 seconds in order to mitigate the values.

The following initial 37 features were considered:

- Mean and variance of: Spectral Centroid, Spectral Rolloff, Spectral Flux, Zero Crossing Rate (ZCR), Short Time Energy (STE), Spectral Flatness Measure (SFM) [22], and Voice2White (V2W) [23].
- High Zero Crossing Rate Ratio (HZCRR), Low Short Time Energy Ratio (LSTER) [24] and percentage of Low-Energy Frames (LEF).
- 20 Mel Frequency Cepstral Coefficients [25].

Since the mean value and the variance for any of the listed features has been considered, the number of features to be selected by the HS algorithm has been found to be $N_F = 2\times37 = 74$. The final 74-feature vector, **F**, is created by calculating these features from both the original time-domain sound signal, and from the linear prediction coefficients (LPC) analysis residual. Note that some of these features have been deliberately chosen because of their high correlation in order to explore the efficiency of the algorithms implemented. As will be shown, the feature selection algorithm has been found to be robust enough to find a suitable solution for this problem.

5.3 Description of the Classifier

In order to evaluate the classification error associated to a subset of features, it is necessary to specify the classifier. In this chapter, the Mean Square Error (MSE) linear classifier has been chosen as one of the most appropriates because of its simplicity and good results.

In a linear classifier, the decision rule depends on C different linear combinations of the input features:

$$g_c = f\left(w_{0c} + \sum_{n=1}^{M} x_n w_{nc} \right) \tag{13}$$

where x_n represents the values of the n-th feature, M represents the number of input features, w_{nc} represents the weights of the linear combination, and g_c is the linear combination obtained for the c-th class (c = 1,..., C). The final decision corresponds to the linear combination with the highest result.

This process can be alternatively described by using matrix notation, defining the input patterns matrix as:

$$\mathbf{Q} = \begin{pmatrix} 1 & 1 & \cdots & 1 \\ x_{11} & x_{12} & \cdots & x_{1N} \\ \vdots & \vdots & \ddots & \vdots \\ x_{M1} & x_{M1} & \cdots & x_{MN} \end{pmatrix} \tag{14}$$

where N represents the number of input patterns. Note that the first row equals 1 in order to implement the independent terms of the linear combinations. The weights of the classifier can defined as:

$$\mathbf{V} = \begin{pmatrix} w_{01} & w_{11} & \cdots & w_{M1} \\ w_{02} & w_{12} & \cdots & w_{M2} \\ \vdots & \vdots & \ddots & \vdots \\ w_{0C} & w_{1C} & \cdots & w_{MC} \end{pmatrix} \tag{15}$$

where C represents the number of classes to classify (C = 3, speech, music, and noise). The output of the linear combinations for the input patterns is:

$$\mathbf{Y} = \mathbf{V} \cdot \mathbf{Q} \tag{16}$$

with $\mathbf{Y}$ being a matrix with C rows and N columns.

The error (E) thus results in being:

$$\mathbf{E} = \mathbf{Y} - \mathbf{T} = \mathbf{V} \cdot \mathbf{Q} - \mathbf{T} \tag{17}$$

where **T** represents a $C \times N$ matrix containing the target classes for each input pattern. If the MSE is assumed to be:

$$MSE = \frac{1}{NC} \sum_{n=1}^{N} \sum_{c=1}^{C} e_{cn}^2 \tag{18}$$

It is possible to derive with respect to the coefficients w_{ij}, and minimize the *MSE*, the result being:

$$\mathbf{V} = \mathbf{T} \cdot \mathbf{Q}^T \cdot \left(\mathbf{Q} \cdot \mathbf{Q}^T\right)^{-1} \tag{19}$$

5.4 Parameters for the Harmony Search Algorithm

In the experiments of this chapter, the HS algorithm has been used in order to determine the set of features that minimizes the validation error rate. For this purpose, several feature sets have been explored by using the optimization process described in Section 3. Table 1 lists the parameters used for the problem at hand. The parameters have been explored to select those more appropriate features for sound classification in digital hearing aids.

Table 1. List of the parameters used for the HS algorithm

Parameter	*Symbol*	*Value*
Number of melodies in memory	M	$M = 20$
Probability of generating a novel melody	P_n	$P_n = 0.1$
Probability of mutation	P_m	$P_m = 0.1$

Regarding the fitness objective, the classification error probability over the validation set has been used. In those cases in which the classification error over the validation set is exactly the same, the mean square error over the validation set has also been used.

Each experiment has been repeated 30 times, and the mean and the standard deviation of the classification error rates have been calculated and evaluated.

6 Results

This section summarizes the results obtained by the HS algorithm in terms of error rate. For comparative purposes, the results applying the sequential search (SS) and random search (RS) algorithms have been also included.

Table 2 shows the results obtained by these feature selection techniques. For the RS and the HS algorithms, three different batches of experiments have been carried out, limiting the number of iterations of the algorithm to 100, 1,000, or 10,000 iterations. These results have been expressed as the mean and the standard deviation for the classification error rates over the validation and the test sets.

As can be seen in Table 2, the SS algorithm obtains the highest test error rate. In this test, this method is deterministic, always obtaining the same solution, as demonstrated by a standard deviation of zero for the classification error rate. The HS algorithm outperformed the RS process, even with only 100 iterations of the algorithm, both in terms of validation error rate and test error rate.

For comparative purposes, the relationships between the number of iterations and the mean test error rate of the HS algorithm and the RS algorithms have been represented in Figure 5. Figure 5 demonstrates the best performance of the HS-based feature selection technique, which can achieve very good results with less than 1,000 iterations.

Table 2. Validation and test classification error rates (%) for different methods

Algorithms	*Iterations*	*Validation error rate (%)*		*Test error rate (%)*	
		Mean	*Std*	*Mean*	*Std*
Sequential search (SS)	-	2.22	0.00	6.18	0.00
Random search (RS)	100	2.69	0.24	5.23	0.70
	1,000	2.26	0.20	5.03	0.65
	10,000	1.85	0.14	4.70	0.63
Harmony search (HS)	100	2.56	0.27	4.82	0.69
	1,000	1.39	0.15	4.31	0.49
	10,000	0.78	0.15	4.14	0.42

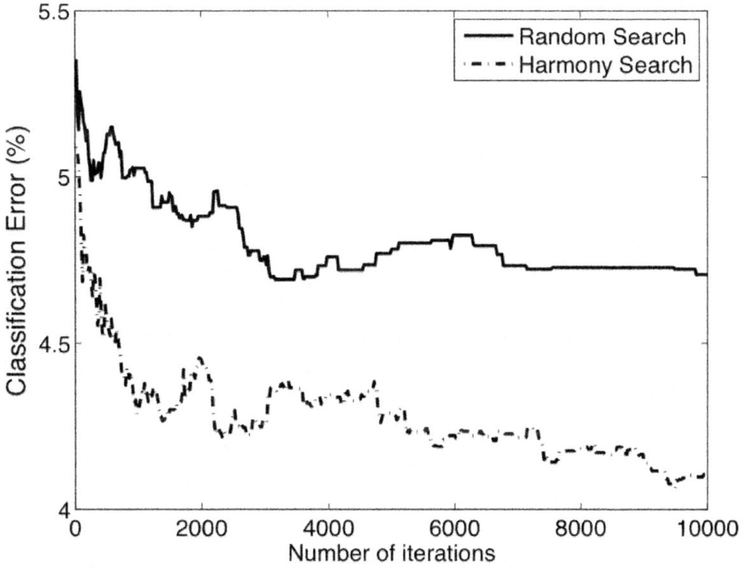

Fig. 5. Relationship between the mean test error rate and the number of iterations for the random search method and the harmony search algorithm

7 Conclusions

This chapter has centered on the novel application of the harmony search algorithm on selecting features for sound classification in hearing aids.

After motivating the social demand for hearing aids exhibiting sound classification, it has been shown how the problem of feature selection in hearing aids is very complicated in the sense that hearing aids suffer from strong constrains that make difficult the design of the application. It has been explained how these limitations arise mainly from the fact that hearing aids must work at very low clock frequency in order to minimize power consumption and maximize battery life. This necessitates to the use of design algorithms that require a small number of operations per second. It has been illustrated that, as a consequence, implementing sound classification algorithms embedded in hearing aids is a very challenging task. The underlying reason is that it requires optimizing each parameter for the sound classifier in order to reduce its computational complexity while maintaining a low error probability, and thus, a good dynamic adaptation to the sound surroundings.

In particular, the feature extraction process, essential to properly characterize the sound to be classified, is arguably one of the most time-consuming tasks. Therefore, reducing the number of features has been shown to be one of the most successful ways of diminishing the complexity of the classification system. However, an excessively small number of features may likely increase the error probability. Thus, selecting those more appropriate features is a crucial point for the application at hand.

The feasibility of the harmony search algorithm for solving this complicated problem has been explored. Harmony search algorithm is an optimization method that mimics the music composition rules in order to search for the best solution to a mathematical problem by minimizing some kind of 'fitness function'. Just in this respect, the first step has been establishing a fitness function based on the classification error. For this purpose, a mean square error linear classifier has been designed for each combination of features, and the classification error rate over the validation set has been used as fitness criterion. So, the harmony search algorithm aims to reduce the number of combinations of features to be evaluated for obtaining the subset of features that minimize this fitness criterion.

The main results obtained by the harmony search algorithm demonstrate the good performance of this feature selection technique. For comparative purposes, those results obtained applying a sequential forward search and a random search process are also included. Considering the same number of iterations of the algorithms, the proposed feature selection technique outperforms these more classical approaches both terms of fitness performance and test error rate.

The harmony search algorithm can reduce the number of feature combinations to be evaluated in the feature selection process for a sound classification system. For the same number of iterations, the final test error rate achieved by the proposed search process is 13% lower than the best test error rate achieved by a random search process. These results make the harmony search algorithm a viable and promising choice in the feature selection problem for sound classification in digital hearing aids.

References

1. Büchler, M.: Algorithms for sound classification in hearing instruments. PhD thesis, Swiss Federal institute of Technology, Zurich (2002)
2. Keidser, G.: The relationships between listening conditions and alternative amplifications schemes for multiple memory hearing aids. Ear Hear 16, 575–586 (1995)
3. Keidser, G.: Selecting different amplification for different listening conditions. J. of the American Academy of Audiology 7, 92–104 (1996)
4. Alexandre, E., Cuadra, L., Álvarez, L., Rosa, M., López, F.: Two-layer automatic sound classification system for conversation enhancement. Integrated Computer Aided Engineering 15, 85–94 (2008)
5. Alexandre, E., Cuadra, L., Rosa, M., Francisco López, F.: Feature selection for sound classification in hearing aids through restricted search driven by genetic algorithms. IEEE Transactions on Audio, Speech and Language Processing 15, 2249–2256 (2007)
6. Alexandre, E., Cuadra, L., Álvarez, L., Rosa-Zurera, M., López-Ferreras, F.: Automatic sound classification for improving speech intelligibility in hearing aids using a layered structure. In: Corchado, E., Yin, H., Botti, V., Fyfe, C. (eds.) IDEAL 2006. LNCS, vol. 4224, pp. 306–313. Springer, Heidelberg (2006)
7. Alexandre, E., Cuadra, L., Rosa, M., López, F.: Speech/non-speech classification in hearing aids driven by tailored neural networks. In: Prasad, B., Prasanna, S.R.M. (eds.) Speech, Audio, Image and Biomedical Signal Processing Using Neural Networks. Springer, Heidelberg (2008)
8. Alexandre, E., Álvarez, L., Cuadra, L., Rosa, M., López, F.: Automatic sound classification algorithm for hearing aids using a hierarchical taxonomy. In: Bialostockiej, W.P., Skarbek, W., Dorbucki, A., Petrovsky, A. (eds.) New Trends in Audio and Video, Bialystok, vol. 1 (2006)
9. Cuadra, L., Alexandre, E., Álvarez, L., Rosa, M.: Reducing the computational cost for sound classification in hearing aids by selecting features via genetic algorithms with restricted search. In: Proceedings of IEEE ICALIP, Shanghai (2008)
10. Alexandre, E., Gil, R., Cuadra, L., Álvarez, L., Rosa, M.: Speech/music/noise classification in hearing aids using a two-layer classification system with MSE linear discriminants. In: Proceedings of European Signal Processing Conference, Switzerland (2008)
11. Amor, J., Alexandre, E., Gil, R., Álvarez, L., Huerta, E.: Music-inspired harmony search algorithm applied to feature selection for sound classification in hearing AIDS. In: Proceedings of AES 124th Convention Proceedings, Amsterdam (2008)
12. Alexandre, E., Cuadra, L., Álvarez, L., Utrilla, M.: Exploring the feasibility of a two-layer NN-based sound classifier for hearing aids. In: Proceedings of European Signal Processing Conference, Poznan (2007)
13. Alexandre, E., Cuadra, L., Álvarez, L., Rosa, M.: NN-based automatic sound classifier for digital hearing aids. In: Proceedings of IEEE WISP, Alcalá de Henares (2007)
14. Alexandre, E., Rosa, M., Cuadra, L., Gil, R.: Application of Fisher linear discriminant analysis to speech/music classification. In: Proceedings of AES 120th Convention, Paris (2006)
15. Duda, R.O., Hart, P.E., Stork, D.G.: Pattern Classification. Wiley-Interscience, Hoboken (2001)
16. Geem, Z.W., Kim, J.H., Loganathan, G.V.: A new heuristic optimization algorithm: harmony search. Simulation 76, 60–68 (2001)

17. Geem, Z.W., Hwangbo, H.: Application of harmony search to multi-objective optimization for satellite heat pipe design. In: Proceedings of US-Korea Conference on Science, Technology & Entrepreneurship, Teaneck (2006)
18. Geem, Z.W., Tseng, C., Park, Y.: Harmony search for generalized orienteering problem: Best touring in china. In: Wang, L., Chen, K., S. Ong, Y. (eds.) ICNC 2005. LNCS, vol. 3612, pp. 741–750. Springer, Heidelberg (2005)
19. Geem, Z.W.: Optimal cost design of water distribution networks using harmony search. Engineering Optimization 38, 259–280 (2006)
20. Geem, Z.W.: Optimal scheduling of multiple dam system using harmony search algorithm. In: Sandoval, F., Prieto, A.G., Cabestany, J., Graña, M. (eds.) IWANN 2007. LNCS, vol. 4507, pp. 316–323. Springer, Heidelberg (2007)
21. Geem, Z.W., Williams, J.C.: Harmony search and ecological optimization. International Journal of Energy and Environment 1, 150–154 (2007)
22. Scheirer, E., Slaney, M.: Construction and evaluation of a robust multifeature speech /music discriminator. In: Proceedings of the IEEE International Conference on Acoustics, Speech, and Signal Processing, Washington, DC (1997)
23. Guaus, E., Batlle, E.: A non-linear rhythm-based style classification for broadcast speech-music discrimination. In: Proceedings of AES 116th Convention, Berlin (2004)
24. Lu, L., Zhang, H., Jiang, H.: Content analysis for audio classification and segmentation. IEEE Transactions on Speech and Audio Processing 10, 504–516 (2002)
25. Davis, S., Mermelstein, P.: Experiments in syllable-based recognition of continuous speech. IEEE Transactions on Acoustics, Speech and Signal Processing 28, 357–366 (1980)

Harmony Search in Therapeutic Medical Physics

Aditya Panchal

Department of Medical Physics, Advocate Lutheran General Hospital,
1700 Luther Lane, Park Ridge, IL 60068, USA
apanchal@bastula.org

Abstract. Medical physics is a branch of physics that concerns the application of radiation for therapeutic and diagnostic use in medicine. In therapeutic medical physics, ionizing radiation is used to treat patients afflicted with cancer. The main goal in radiotherapy is to deliver a high amount of radiation dose to cancer cells while sparing surrounding organs. The process begins with radiation treatment planning that requires optimization of radioisotope placement or radiation beam intensities. Prior optimization algorithms, such as genetic algorithm and simulated annealing, have been widely discussed in medical physics literature. However, Harmony Search has shown to be a superior optimization algorithm based on the results in other scientific fields. Thus, this study investigated high dose-rate prostate brachytherapy optimization using Harmony Search. Results showed that the algorithm is significantly faster than genetic algorithm and that rapid planning allows improved patient care in medical physics.

Keywords: Medical Physics, Radiation Oncology, Brachytherapy, Optimization, Harmony Search.

1 Introduction

Radiation therapy has been used to treat cancer for over a hundred years. In that time, significant advances have been made in both the localization of cancer and the delivery of radiation. Throughout the last decade, more precise targeting of tumors due to new treatment modalities have allowed customization of the radiation dose which in turn minimizes dose to adjacent critical structures and ensures that the prescribed dose is delivered to the tumor. Some of these modalities include Intensity Modulated Radiation Therapy (IMRT) and high dose-rate (HDR) brachytherapy.

The purpose of therapeutic radiation stems from the radiobiological effect of ionizing radiation interacting with the DNA of cancer cells. This renders them frozen in a specific stage of the cell cycle, unable to reproduce. Since the radiation dose prescription can be quite high, it is usually necessary to divide the treatment into multiple fractions. These treatments are then performed until the total dose is delivered. This enables healthy normal tissue to regenerate between fractions and also maximizes the damage to tumor cells.

In order to deliver the radiation to a patient, the interaction of the particles must be modeled on a radiation treatment planning system. Since these new modalities allow significant choice in how to deliver the treatment, optimization of the radiation beam intensity patterns (in IMRT) or radioactive source placement (in HDR brachytherapy) must be performed to generate an effective plan for each individual patient. In the past, optimization algorithms, such as evolutionary algorithms (including genetic algorithm) [1], simulated annealing [2] and linear programming [3],

Z.W. Geem (Ed.): Music-Inspired Harmony Search Algorithm, SCI 191, pp. 189–203.
springerlink.com

have been used. Although these algorithms are suitable for optimization, they still require a significant amount of time to generate the most optimal treatment plan. In the medical field, quick intervention can be the key to the improvement of cancer control rates. Thus, shortening the time spent generating a plan can allow the patient to be treated sooner.

As a result of reviewing the remarkable results demonstrated in other scientific fields [4-9], Harmony Search was chosen to tackle the problem of optimizing HDR brachytherapy for prostate cancer. Radiation dose-based optimization constraints were evaluated in order to determine the most optimal treatment plan. Based on the results of this investigation, Harmony Search is considered a natural fit for therapeutic medical physics, which can be applied to similar scenarios within the field.

2 Radiation Treatment Planning

The typical process of delivering radiation to a patient involves acquiring a set of images of the patient, such as a computed tomography (CT) scan. Prior to the scan, the patient is placed in the position that they are to be treated in so that the images reflect reproducible anatomy and geometry. Next, the scan is imported into the radiation treatment planning system. This sophisticated system is then able to reconstruct the patient in 3D, allow the radiation target and critical structures to be delineated, and to place, calculate and evaluate radiation dosimetry.

The radiation oncologist will first decide what part of the anatomy will be treated and contour those structures on the image scan. Additionally, the physician will specify the prescription dose the target must receive. After this is completed, the organs-at-risk (OARs) are outlined. The next step is to optimize the parameters which control the radiation, virtually simulate the radiation interactions in the patient and determine the final planned dose to the target and the OARs. The plan can then be evaluated by using dose-volume histograms.

2.1 Overview of Dose-Volume Histograms

The dose volume histogram (DVH) was described by Chen et al. in 1987 [10]. It is a version of the standard histogram that allows the analysis of a tabulated data set, with regards to the frequency each data point occurs. The data is separated into bins or ranges, and each time a data point falls into a bin, the frequency of that bin is increased by one. The frequency of data is plotted against the bin value.

In the dose volume histogram, dose values (typically measured in cGy or Gy) are tabulated either by points or voxels within a specific organ structure. Each structure is calculated separately and the plot is shown as the volume of the structure receiving a given dose or higher as a function of dose. This form of the DVH is known as a cumulative DVH, since the plot shows the volume that receives a specific dose or higher. The bin value can be set to any desired dose interval, which will modify the coarseness of the DVH. An example of a DVH with a bin value of cGy is given in Figure 1.

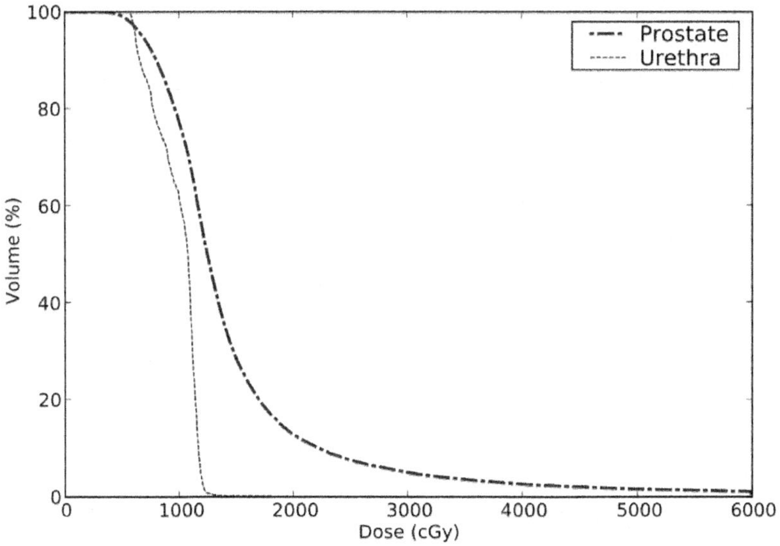

Fig. 1. A cumulative dose-volume histogram (DVH)

Once the DVH is created, it can be analyzed by specifying two types of criteria: volume and dose constraints. These parameters are standard nomenclature in the field of radiation oncology. The first letter determines the type of constraint (dose or volume) and the following number identifies which value to look up from the DVH. For example, if a physician would like to determine the V100 of the prostate, he would like to know the volume of the prostate that receives 100% of the prescribed dose. On the other hand, if the D90 is requested, the dose that 90% of the volume receives is returned. Additionally, volume constraints (Vxxx) can also be specified in absolute volume, as well as percent of the total volume. In this version of the constraint, the standard notation is to append a "cc" after the number, i.e. D1cc (determine the dose to 1 cc of the volume).

3 High Dose-Rate Prostate Brachytherapy

Brachytherapy is a form of radiation therapy in which sealed radioactive source(s) are placed inside or on the body to irradiate tumors. With improved anatomical resolution, physicians are able to define target volumes more clearly with image guidance. Brachytherapy can be used to treat these targets in a variety of situations such as interstitial, intracavitary, intravascular or even on the surface of tumor.

In high dose-rate (HDR) prostate brachytherapy, approximately 15 to 25 catheters are inserted into the target volume for treatment and a single source (commonly ^{192}Ir) is moved to dwell positions within the catheters. Each position to which the source can be advanced is known as a dwell position. The source remains in each position for a specified amount of time. For each catheter, a step distance can be defined, in which the source is retracted by the given distance in order to deliver a desired dose distribution. It is this combination of the remote afterloader mechanism and the HDR source that allows for tight control of the dose delivery.

A computer simulation is performed in order to get an optimal dose distribution that defines the dwell time or weighting of each dwell position. Once obtained, the planned dose to the target and the OARs can be evaluated by using DVHs.

A desired dose distribution for a HDR prostate brachytherapy treatment is typically generated by examining a 3D CT image set of a patient with catheters implanted into the prostate. Optimization of the dwell position and time is performed by trial and error methods by a medical dosimetrist or medical physicist. There are a number of documented methods that suggest optimization of treatment plans using DVH-based criteria [11-13].

With advancements in computing power, brachytherapy treatment planning systems can optimize a large number of dwell positions in order to achieve a particular user-specified dose distribution. This method, known as inverse planning, enables the user to specify the constraints, while the planning system manipulates the decision variables in order to satisfy these criteria.

Radiation dosimetry for HDR prostate brachytherapy currently follows certain guidelines. The Radiation Therapy Oncology Group (RTOG) has published dosimetric guidelines for clinical trials for HDR prostate brachytherapy treatments. RTOG Protocol 0321 [14] states:

- The target volume is delineated as the prostate for early stage cases
- Prescription dose is 19 Gy delivered over two fractions to the periphery of the target
- Prescription goal is to deliver 100% of the prescription dose to 90% of the target (D90 $\leq$ prescription dose)
- Less than 1 cc of the bladder and rectum receive 75% of the prescription dose (V75 < 1 cc)
- Less than 1 cc of the urethra receives 125% of the prescription dose (V125 < 1 cc)

As can be seen, the goal of meeting the above dosimetric criteria is accomplished by modifying the relative dwell times or weights of each dwell position of the radioactive seed.

3.1 DVH-Based Objective Function and Optimization Constraints

A general form of a DVH-based objective function can be written as:

$$f = \sum_i c_i \left\{ \sum_k u_{ik} (d_i(V_{i,k}) - D_{i,V_k})^2 \right\} \tag{1}$$

where, for the ith structure, $d_i(V_{i,k})$ and $D_{i,Vk}$ are the kth calculated and prescribed dose-volume constraint for the structure, u_{ik} is the weight assigned to that particular dose-volume constraint, and c_i is the overall importance factor for this structure [11]. This function can be modified for additional constraints such as the maximum dwell time or the maximum dwell position for a given catheter.

In this investigation, Equation 1 was modified to the specific constraints relevant to the simulation:

$$f = 5\times(90 - D100_{prostate})^2 + (1 - V125_{urethra})^2 + (1 - V75_{bladder})^2 + (1 - V75_{rectum})^2 \tag{2}$$

This modification takes into account both dose and volume constraints for the prostate, urethra, bladder and rectum from RTOG Protocol 0321 shown in Table 1.

Table 1. DVH-based objective function constraints based on RTOG Protocol 0321

i	Volume	Constraint, $D_{i,Vk}$	Expected Value, $d_i(V_{i,k})$
1	Prostate	D100	90%
2	Bladder	V125	1 cc
3	Rectum	V75	1 cc
4	Urethra	V75	1 cc

All constraints were weighted equally, with exception of the prostate, which was set to 5. The increased importance of the prostate reflects the fact that it is the target volume. The prescription dose was set to 950 cGy, also from the RTOG protocol.

3.2 Optimization Parameters

The simulation environment was constructed using various open source software including: Python, wxPython, NumPy, and Matplotlib. In the environment, the user is

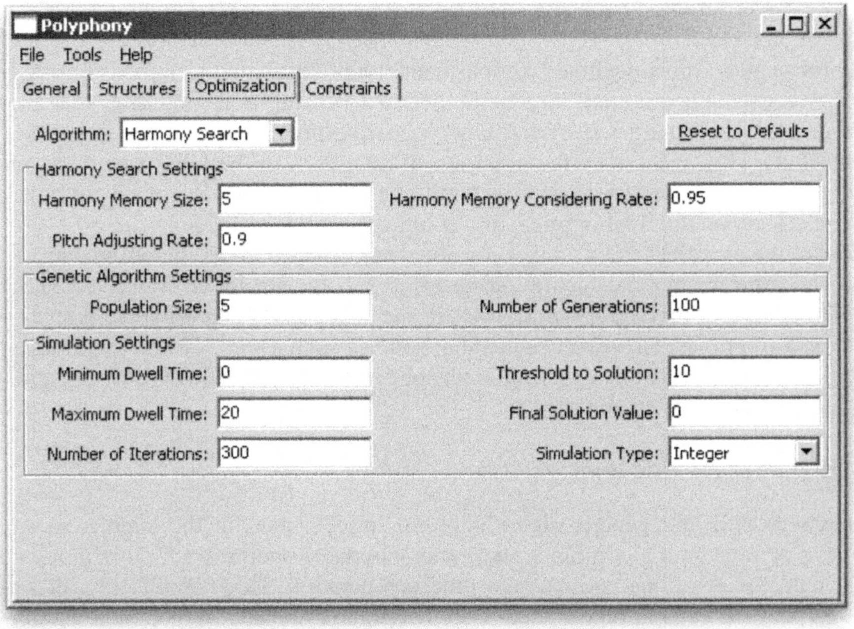

Fig. 2. Default optimization settings for the simulation

able to choose either Harmony Search or genetic algorithm as the optimization algorithm. A Windows XP PC with an Intel Core 2 Duo E6300 Processor (2 processor cores at 1.86 GHz) and 2 GB of RAM was used for all of the simulations. The default settings for the simulation can be seen in Figure 2.

For Harmony Search, the default values for the harmony memory size (HMS), harmony memory considering rate (HMCR), and pitch adjusting rate (PAR) were 5, 0.95, and 0.9 respectively and could be modified in the simulation. For the genetic algorithm, the population size (default value of 5) and the number of generations (default value of 100) could be adjusted.

The general simulation parameters include the minimum and maximum dwell times, which were set as hard constraints. At each dwell position, the dwell times were constrained to lie between 0 and a maximum of 20 seconds for both algorithms. The threshold to solution of the objective function was set to 10.

Additionally, the final solution value was set to 0. With this combination, the simulation would terminate if the solution value were less than 10. The number of iterations was set to 300, in order to let each simulation reach the solution. This number was determined from prior experimentation as a minimum threshold, which would allow the simulation to reach the goal. Increasing the number of iterations increases the total time for the simulation, but does not necessarily produce a better solution.

Finally, the simulation could be operated in integer mode or floating point mode. In integer mode, the dwell times are constrained to integer values as mentioned above. In floating point mode, the dwell times are represented in computer hardware as binary fractions, and is the method how real numbers are stored. For example, the decimal fraction 0.125 has a value of 1/10 + 2/100 and 5/1000 while the binary fraction 0.001 has a value of 0/2 + 0/4 + 1/8. These two fractions have identical values, but the only difference is the first value is written in base 10 notation and the second is written in base 2.

Unfortunately, most decimal fractions cannot be represented exactly as binary fractions. A consequence is that, in general, decimal floating point numbers that are entered are only approximated by the binary floating-point numbers actually stored in the machine. However, since the precision of time used in the HDR treatment unit is either specified as an integer or to the tenth of a decimal point, this does not have any adverse effects on the final optimization solution.

These modes reflect the data type for the dwell times used in the optimization. For all of the experiments, except for subsection 3.3.2, integer mode was selected, since the GammaMed ^{192}Ir HDR source (used in our institution) can only use integer seconds for dwell times. However, since other sources such as the Varian VariSource can be operated with tenth of a second precision, it is important to investigate how the optimization differs between modes.

3.3 Results of Investigation

Harmony Search and genetic algorithm were implemented in the simulation as the available optimization algorithms. They were compared against each other in regard to the number of iterations and average time per iteration. Additionally, integer versus floating point simulation modes were also investigated. Finally, the optimal parameters for Harmony Search were determined for HDR prostate brachytherapy optimization.

Table 2. Comparison of Harmony Search and genetic algorithm optimization - average number of iterations and average total simulation time

Patient	Optimization Method	Iterations	Simulation Time (s)
1	Harmony Search	116	2515
	Genetic Algorithm	403	7870
	% Difference	**347%**	**313%**
2	Harmony Search	48	940
	Genetic Algorithm	254	4388
	% Difference	**529%**	**467%**
3	Harmony Search	25	474
	Genetic Algorithm	149	2313
	% Difference	**606%**	**488%**
4	Harmony Search	33	437
	Genetic Algorithm	178	2267
	% Difference	**536%**	**518%**
5	Harmony Search	62	210
	Genetic Algorithm	255	1200
	% Difference	**411%**	**571%**
6	Harmony Search	150	438
	Genetic Algorithm	473	1036
	% Difference	**315%**	**236%**
7	Harmony Search	40	935
	Genetic Algorithm	154	3793
	% Difference	**385%**	**406%**
8	Harmony Search	13	64
	Genetic Algorithm	33	113
	% Difference	**254%**	**176%**
9	Harmony Search	40	129
	Genetic Algorithm	130	459
	% Difference	**325%**	**355%**
Average	**% Difference**	**412%**	**392%**

3.3.1 Comparison of Harmony Search and Genetic Algorithm Optimization

Harmony Search and genetic algorithm were used to optimize nine patients using the constraints specified in Table 1. Each algorithm run was repeated five times for each patient in order to improve statistical significance. During each simulation, the optimizer was allowed to run completely, such that both algorithms would reach the same end result. The results of the comparison are presented in Table 2.

It should be noted that the average number of iterations in Harmony Search did not include the iterations used to construct the harmony memory (HM). Likewise, the average number of iterations in genetic algorithm did not include the iterations used to generate the initial population. The results show that Harmony Search, in the best case, is at least 5 times faster and in the worst case 3 times faster when compared to the genetic algorithm. On average, about a 400% improvement can be realized by using Harmony Search over the genetic algorithm as the optimization algorithm.

The reason that Harmony Search is so much faster than genetic algorithm can be attributed to the fact that the former selects values from all available vectors in the Harmony Memory. The genetic algorithm chooses values only from two vectors. Both algorithms however, do have the ability to introduce variability into each decision variable by replacing a value at random (Harmony Memory Considering Rate / Pitch Adjusting Rate for the Harmony Search and crossover / mutation for the genetic algorithm).

These results are in line with prior studies [5-9]. The reason that the optimization is slower with the DVH-based objective function compared to simple mathematical cases of finding the solution to the "six hump camel back" function [15], is that the dose calculation involves much more decision variables (dwell positions) versus only 2 decision variables. In all of the patients listed above, the number of dwell positions was at least 200 and was sometimes over 400. This number was dependent on the specific patient data imported, where a shorter step distance involves more possible dwell positions.

A note of interest is the average time per iteration. In Table 3, a selection of patients is shown with the corresponding time per iteration values. Although the genetic algorithm is slower to reach the convergence point, the time per iteration is slightly faster. The reason is that, for each iteration, Harmony Search must assemble a vector from the Harmony Memory and compare it once per iteration, while the genetic algorithm only has to select two individuals, create an offspring and create a new population once per generation.

Table 3. Comparison of Harmony Search and genetic algorithm optimization - average time per iteration

Patient	Optimization Method	Average Time / Iteration (s)
2	Harmony Search	21.68
	Genetic Algorithm	19.53
3	Harmony Search	19.58
	Genetic Algorithm	17.28
4	Harmony Search	19.28
	Genetic Algorithm	15.53
5	Harmony Search	6.84
	Genetic Algorithm	6.32
Average	% Difference	11.34%

For example, if the Harmony Memory Size was set to 5, then, for each dwell position, the algorithm must select a value either from the 5 vectors in memory or choose a random value. If the value is selected from memory, the algorithm also must vary the pitch, according to the Pitch Adjusting Rate. In this simulation, that would be equivalent to increasing or decreasing the dwell time by one. This whole process must be repeated for each dwell position. After this, the new vector has to be compared against the existing vectors in the Harmony Memory. If it is better, it replaces the vector with the worst function value.

On the other hand, the genetic algorithm selects two individuals (solutions) to mate. The dwell times in each solution are crossed over and potentially mutated in order to produce a new, but similar solution. Only after the mating process has been completed does the comparison step occur. It is this difference that causes Harmony Search to be somewhat slower in speed per iteration. However, due to the significant convergence speed difference, the iteration speed becomes insignificant after all.

3.3.2 Integer Versus Floating Point Dwell Time Optimization

Since the optimization can be run in both integer and floating point (FP) mode, it is important to investigate the difference between them during the simulation. Patient #4 was selected and optimized 10 times each for Harmony Search using both integer and floating point modes. The same experiment was repeated for the genetic algorithm. The constraints for this simulation were set to the same values as in Table 1.

As can be seen in Table 4, Harmony Search takes slightly longer (average number of iterations and average time per iteration) to converge to the solution when using floating point mode compared to integer mode. However, the genetic algorithm actually is slightly faster when using floating point mode.

Table 4. Comparison of Harmony Search and genetic algorithm optimization - Integer versus Floating Point

Optimization Method	Integer / FP	Iterations	Time / Iteration (s)
Harmony Search	Integer	56.70	2.84
	FP	69.33	3.19
Genetic Algorithm	Integer	198.20	3.21
	FP	182.43	3.18

If more iterations were performed for each type of simulation, the difference would probably be negligible and become statistically insignificant. Still, Harmony Search is faster than the genetic algorithm for both floating point and integer modes for each overall simulation, corroborating the results shown above in subsection 3.3.1. In integer mode, Harmony Search is around 400% faster and in floating point mode almost 300% faster.

The DVHs for integer and floating point representations using Harmony Search are shown in Figure 3 and Figure 4, respectively. Qualitatively, Harmony Search in floating point mode meets the same constraints as the integer mode, however, the dose to the prostate and urethra are much higher. Likewise, similar results occur with the genetic algorithm in floating point mode. The reason is that there are only 21 possible choices for the integer times (0-20), while there are infinitely more choices when using floating point. In order for the floating point simulation to minimize the remaining dwell times, it would take significantly more time to reduce to an integer mode-like result.

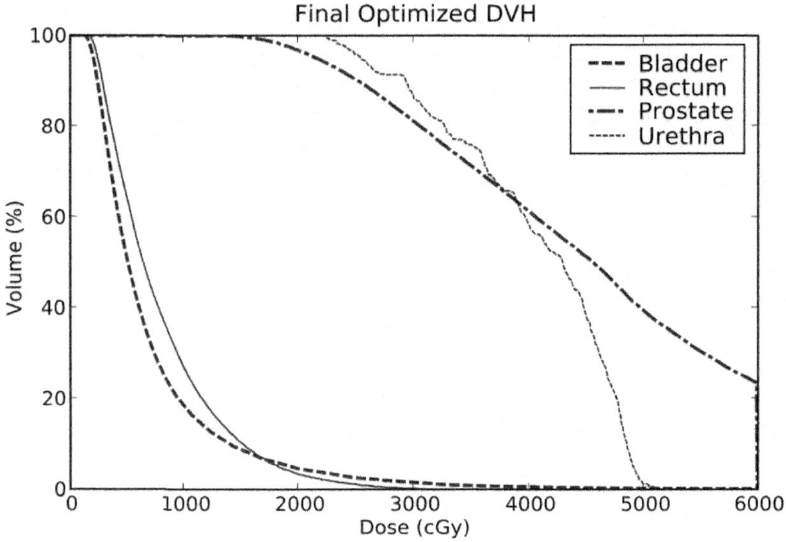

Fig. 3. DVH for Harmony Search in integer mode

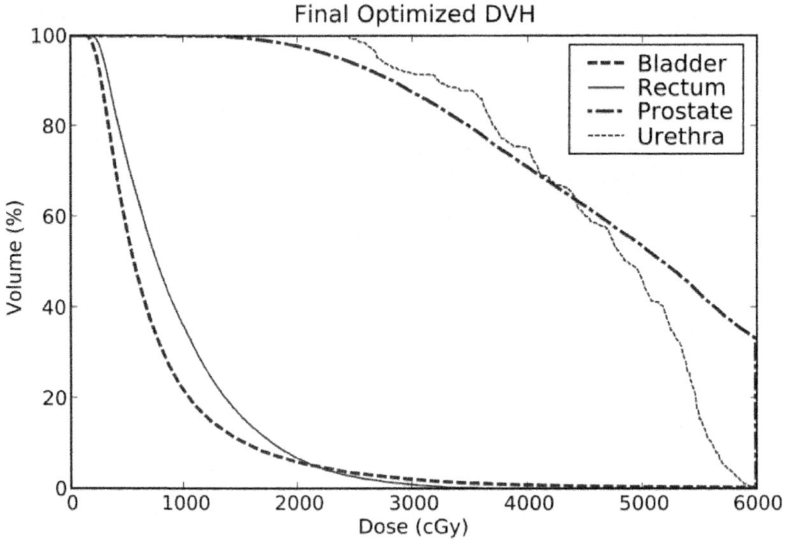

Fig. 4. DVH for Harmony Search in floating point mode

For HDR brachytherapy treatment units that only support integer dwell times, it would make sense to optimize only in integer mode. Not only is it slightly faster (for Harmony Search), but also produces a better result for the optimization. However, if the optimization for floating point mode were continued with more iterations and the constraints were pushed harder, the dose to the prostate and urethra would fall.

3.3.3 Harmony Search Parameters

When using Harmony Search, a number of parameters can be modified in order to change the outcome of the optimization. The Harmony Search parameters: HMS, HMCR, and PAR were modified independently in order to determine the best values for HDR prostate brachytherapy optimization.

For each parameter investigation, the number of iterations was set to 300. Additionally, the constraints were modified so that the prostate D95 was set to 100% and the urethra, rectum, and bladder weights were increased to 50. This was done to speed up the convergence time due to repeated simulations. For each Harmony Memory Size, the experiment was repeated five times to improve statistical significance.

3.3.3.1 Harmony Memory Size. The Harmony Memory Size is an analogous construct to the population size of the genetic algorithm. Typical values from previous studies range from 4 to 10 [5-9].

A new patient (#10) was selected to run the optimization and the Harmony Memory Size was set to 1, 5, 10, and 20. The results are shown in Table 5. The results show that as the Harmony Memory Size increases in value, the convergence of the solution gets worse, based on the final solution value. Additionally, the time each iteration takes increases as well. The reasoning behind this is that during each iteration, the algorithm selects randomly from the Harmony Memory to construct a new vector. If the Harmony Memory Size increases, the chance of a better solution to be chosen decreases since the HM is populated with more, but inferior solutions.

Table 5. Harmony Memory Size (HMS) comparison

HMS	Solution Value	Time / Iteration (s)	$D95_{prostate}$ (cGy)
1	689 ± 65	6.58 ± 0.64	947.0 ± 3.46
5	838 ± 57	6.91 ± 0.62	954.0 ± 4.24
10	1104 ± 145	7.50 ± 0.86	948.0 ± 0.00
20	1536 ± 168	9.95 ± 1.03	944.0 ± 5.13

When the Harmony Memory Size is set to a value of 1, the convergence is the fastest. At first, it may seem that this condition may result in premature convergence in a local minimum / maximum. However, Harmony Search introduces random values based on the Harmony Memory Considering Rate, and therefore, decreases this possibility since it allows to escape the current minimum / maximum.

Sample DVHs were constructed for each of the HMS values chosen. The difference between each DVH is minimal, but if inspected closely, and correlated with the solution value, the DVH from HMS sizes 1 and 5 are slightly better than 10 and 20 as the urethra receives less of a dose than the prostate.

3.3.3.2 Harmony Memory Considering Rate. The Harmony Memory Considering Rate (HMCR) is a variable that determines whether the value for the current decision variable in the new vector should come from Harmony Memory or be randomly generated. This allows variability so that the optimization does not get trapped in a local minimum maximum. Prior studies have used a value of 0.95 for the HMCR.

Patient #10 was selected once again to run the optimization and the Harmony Memory Considering Rate was set to 0.3, 0.5, 0.7, and 0.95. For each HMCR value, five runs were completed to improve statistical significance. The results are shown in Table 6.

Table 6. Harmony Memory Considering Rate (HMCR) comparison

HMCR	Solution Value	Time / Iteration (s)	$D95_{prostate}$ (cGy)
0.3	2862 ± 153	3.91 ± 0.54	783.0 ± 15.02
0.5	2387 ± 129	4.68 ± 0.62	921.0 ± 13.08
0.7	1363 ± 145	5.02 ± 0.58	933.0 ± 7.76
0.95	911 ± 133	5.31 ± 0.74	951.0 ± 6.00

As can be seen in Table 6, the Harmony Memory Considering Rate improves the rate of convergence for the optimization. This makes sense intuitively, since the best vectors discovered are stored within the Harmony Memory. The average time per iteration between HMCR values shows that as decision variable values are taken at random from the Harmony Memory, it takes more time to look them up. This is further complicated by the fact that after a value is selected from HM, it has a chance to enter the pitch adjusting loop.

3.3.3.3 Pitch Adjusting Rate. The Pitch Adjusting Rate (PAR) is a variable similar to genetic algorithm's mutation rate. Typical values from previous studies range from 0.3 to 0.99 [5-9].

Patient #10 was selected once more to run the optimization and the Pitch Adjusting Rate was set to 0.3, 0.5, 0.7, and 0.9. Like the experiments for the Harmony Memory Size and the Harmony Memory Considering Rate above, for each PAR value, five runs were completed to improve statistical significance. The results are shown in Table 7.

Table 7. Pitch Adjusting Rate (PAR) comparison

PAR	Solution Value	Time / Iteration (s)	$D95_{prostate}$ (cGy)
0.3	916 ± 202	4.17 ± 0.43	938.0 ± 35.07
0.5	871 ± 118	4.43 ± 0.57	945.0 ± 5.37
0.7	854 ± 110	4.94 ± 0.51	947.0 ± 5.77
0.9	798 ± 111	5.31 ± 0.76	951.0 ± 5.74

It would be assumed that by decreasing the PAR, the variability for each new vector would decrease, thus decreasing the final solution value. This seems to hold true when the PAR is decreased from 0.9. Not surprisingly, the average time per iteration increases as the PAR increases. This is due to the fact that in the implementation of Harmony Search, a random value must be chosen and multiplied by the pitch distance value. This occurs 90% of time for each decision variable, if the PAR is set to 0.9. The pitch adjusting loop will be mostly bypassed if the PAR is set to a very low value. Thus, decreasing the value of the PAR will not only decrease the probability of pitch adjustment, but also slow down the convergence time.

4 Intensity Modulated Radiation Therapy

In external beam radiotherapy, a linear accelerator is used to accelerate electrons of energies between 6 and 23 MeV, which hit a tungsten target in order to produce ionizing photon radiation. These photon beams, in turn, are collimated down via adjustable apertures that are precisely shaped for a tumor within a patient. The linear accelerator has a rotating gantry that is able to treat a patient using an isocentric technique. Typically, a fixed number (anywhere from 5-9) of beam angles are used to target the tumor so that it is irradiated from different directions in order to decrease toxic radiation dose to surrounding critical structures.

Over the last decade, multileaf collimators (MLC) have been introduced which allow the collimation to be modulated in real time during the beam delivery process. This treatment is more commonly known as Intensity Modulated Radiation Therapy (IMRT). Instead of a single dose of radiation delivered through a shaped aperture, IMRT allows a checkerboard-like intensity pattern of dose to be treated. Each section of the beam that has a different intensity is considered a beamlet. Depending on how long the MLC leaves remain in the path of the radiation will modify how much dose is delivered directly beneath the blocked area. An example of an intensity pattern from a posterior beam entry for a prostate cancer patient is given in Figure 5.

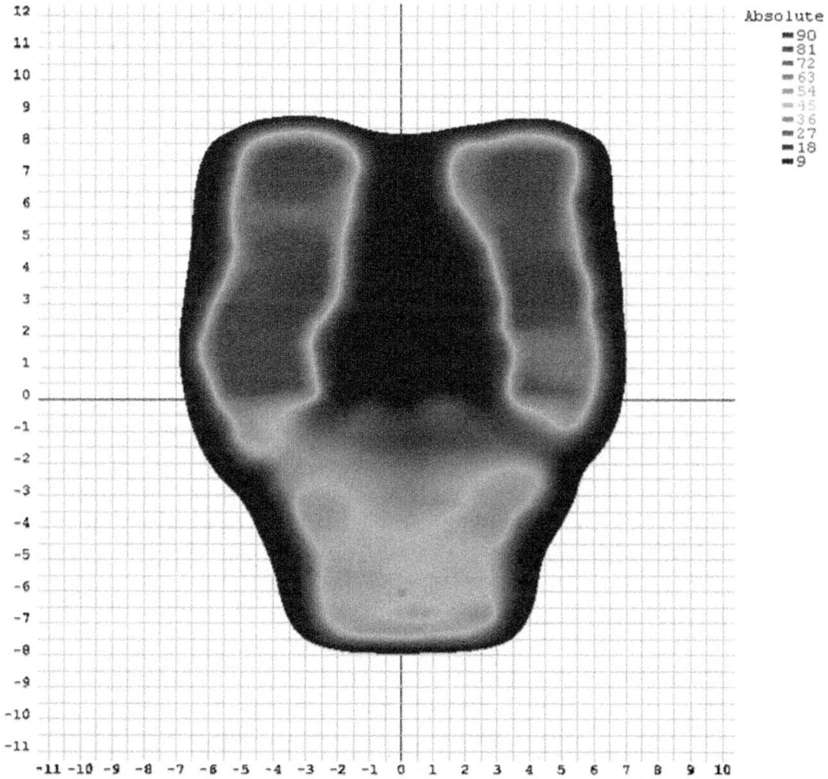

Fig. 5. IMRT intensity pattern from a posterior entry beam for a prostate cancer patient

Success in meeting the radiation amount prescribed to the tumor by a physician, while meeting critical organ tolerances, requires modification of the intensities to be delivered through each aperture. Due to the time consuming nature of modifying each beamlet intensity value, optimization is inherently required for IMRT treatment planning. Current commercial treatment planning systems use simulated annealing to optimize each beamlet and can take anywhere from 30 minutes to 1 hour to meet the prescription criteria.

Based on the results in brachytherapy as discussed above, the introduction of Harmony Search to the IMRT would provide a significant reduction in the time spent in treatment planning. Additionally, a modification of the IMRT technique, known as Volumetric Modulated Arc Therapy (VMAT), has been recently introduced in radiation therapy departments across the globe, which modulates the intensities while the gantry is rotating [16]. This would increase the possible decision variables (beamlet intensities) anywhere from 5- to 10-fold as current IMRT treatments only use 5-9 beam angles, while rotational therapy can be modeled as approximately 50-60 discrete beam angles.

Although the promise of VMAT would allow increased dose delivery to the tumor and decreased dose to critical structures, the significant increase of computation time to plan such a treatment is a perfect fit for Harmony Search to be applied to this new technique in medical physics.

5 Conclusions

This chapter reviewed the novel application of Harmony Search as an optimization algorithm to the field of medical physics. A DVH-based objective function was created and used for the optimization simulation in HDR brachytherapy for prostate cancer. Harmony Search and genetic algorithm were employed as optimization algorithms for the simulation and were compared against each other for nine different patients. The comparison between Harmony Search and genetic algorithm showed that Harmony Search was over four times faster when compared over multiple data sets. The average time per iteration was found to be faster for the genetic algorithm than for Harmony Search due to the fact that the latter must randomly choose decision variable values from the Harmony Memory to create a new solution vector once per iteration; the former chooses only two parents per iteration to create a new solution vector. Additionally, using floating point values for dwell times, Harmony Search was still at least four times faster than the genetic algorithm, corroborating the results for the integer mode. However, in floating point mode, the simulation produces less than satisfactory results and unless required, integer mode should be used. Since the GammaMed treatment unit uses integer dwell times, integer mode is a perfect match. Finally, the optimal values for the Harmony Search parameters (HMS, HMCR and PAR) were determined for the HDR prostate brachytherapy simulation.

In conclusion, Harmony Search was shown to be quite capable as an optimization algorithm for the use in HDR prostate brachytherapy. It is a suitable alternative to existing algorithms. Coupled with the optimal parameters for the algorithm and a multithreaded simulation, this combination has the capability to significantly decrease the time on time-intensive clinic problems such as brachytherapy, IMRT, VMAT, TomoTherapy, and beam angle optimization.

References

1. Schreibmann, E., Lahanas, M., Xing, L., Baltas, E.: Multiobjective evolutionary optimization of the number of beams, their orientations and weights for intensity-modulated radiation therapy. Phys. Med. Biol. 49, 747–770 (2004)
2. Lessard, E., Pouliot, J.: Inverse planning anatomy-based dose optimization for HDR-brachytherapy of the prostate using fast simulated annealing algorithm and dedicated objective function. Medical Physics 28, 773–779 (2001)
3. Jozsef, G., Streeter, O.E., Astrahan, M.A.: The use of linear programming in optimization of HDR implant dose distributions. Medical Physics 30, 751–760 (2003)
4. Geem, Z.W., Kim, J.H., Loganathan, G.V.: A new heuristic optimization algorithm: harmony search. Simulation 76, 60–68 (2001)
5. Kim, J.H., Geem, Z.W., Kim, E.S.: Parameter estimation of the non-linear Muskingum model using harmony search. Journal of the American Water Resources Association 37, 1131–1138 (2001)
6. Geem, Z.W., Kim, J.H., Loganathan, G.V.: Harmony search optimization: application to pipe network design. International Journal of Modeling & Simulation 22, 125–133 (2002)
7. Lee, K.S., Geem, Z.W.: A new structural optimization method based on the harmony search algorithm. Computers and Structures 82, 781–798 (2004)
8. Geem, Z.W., Lee, K.S., Park, Y.: Application of harmony search to vehicle routing. American Journal of Applied Sciences 2, 1552–1557 (2005)
9. Geem, Z.W., Choi, J.: Music composition using harmony search algorithm. In: Giacobini, M. (ed.) EvoWorkshops 2007. LNCS, vol. 4448, pp. 593–600. Springer, Heidelberg (2007)
10. Chen, G.T., Pelizzari, C.A., Spelbring, D.R., Awan, A.: Evaluation of treatment plans using dose volume histograms. Front Radiat. Ther. Oncol. 21, 44–55 (1987)
11. Chen, Y., Boyer, A.L., Xing, L.: A dose-volume histogram based optimization algorithm for ultrasound guided prostate implants. Medical Physics 27, 2286–2292 (2000)
12. Kemmerer, T., Lahanas, M., Baltas, D., Zamboglou, N.: Dose-volume histograms computation comparisons using conventional methods and optimized fast Fourier transforms algorithms for brachytherapy. Medical Physics 27, 2343–2356
13. Karouzakis, K., Lahanas, M., Milickovic, N., Giannouli, S., Baltas, D., Zamboglou, N.: Brachytherapy dose–volume histogram computations using optimized stratified sampling methods. Medical Physics 29, 424–432 (2002)
14. Hsu, I.C., Shinohara, K., Pouliot, J., Purdy, J., Michaelski, J., Ibbot, G.: RTOG protocol 0321 - phase II trial of combined high dose rate brachytherapy and external beam radiotherapy for adenocarcinoma of the prostate (2006) (accessed October 31, 2008), `http://www.rtog.org/members/protocols/0321/0321.pdf`
15. Csendes, T., Ratz, D.: Subdivision direction selection in interval methods for global optimization. SIAM Journal on Numerical Analysis 34, 922–938 (1997)
16. Otto, K.: Volumetric modulated arc therapy: IMRT in a single gantry arc. Medical Physics 35, 310–317 (2008)

Author Index

Printed by Printforce, the Netherlands